PRENTICE-HALL FOUNDATIONS
OF MODERN BIOLOGY SERIES

WILLIAM D. MCELROY
AND CARL P. SWANSON, *editors*

BATES   *Man in Nature, 2nd edition*

BOLD   *The Plant Kingdom, 2nd edition*

BONNER AND MILLS   *Heredity, 2nd edition*

DETHIER AND STELLAR   *Animal Behavior, 2nd edition*

GALSTON   *The Life of the Green Plant, 2nd edition*

HANSON   *Animal Diversity, 2nd edition*

MCELROY   *Cell Physiology and Biochemistry, 2nd edition*

SCHMIDT-NIELSEN   *Animal Physiology, 2nd edition*

SUSSMAN   *Growth and Development, 2nd edition*

SWANSON   *The Cell, 3rd edition*

WALLACE AND SRB   *Adaptation, 2nd edition*

WHITE   *Chemical Background for the Biological Sciences*

THE
CELL

3rd edition

CARL P. SWANSON
The William D. Gill Professor of Biology, The Johns Hopkins University

PRENTICE-HALL, INC.

ENGLEWOOD CLIFFS, NEW JERSEY

TO D. N. S.,
FOR MANY
REASONS

FOUNDATIONS OF MODERN BIOLOGY SERIES   WILLIAM D. MCELROY
AND CARL P. SWANSON, *editors*

C—13-121681-3
P—13-121673-2
Library of Congress Catalog Card Number   69–10438

Current printing   *10 9 8 7 6 5 4 3 2*

PRENTICE-HALL INTERNATIONAL, INC., *London*
PRENTICE-HALL OF AUSTRALIA, PTY. LTD., *Sydney*
PRENTICE-HALL OF CANADA, LTD., *Toronto*
PRENTICE-HALL OF INDIA PRIVATE LTD., *New Delhi*
PRENTICE-HALL OF JAPAN, INC., *Tokyo*

THIS SERIES, FOUNDATIONS OF MODERN BIOLOGY, WHEN LAUNCHED A number of years ago, represented a significant departure in the organization of instructional materials in biology. The success of the series provides ample support for the belief, shared by its authors, editors, and publisher, that student needs for up-to-date, properly illustrated texts and teacher prerogatives in structuring a course can best be served by a group of small volumes so planned as to encompass those areas of study central to an understanding of the content, state, and direction of modern biology. The twelve volumes of the series still represent, in our view, a meaningful division of subject matter.

This edition thus continues to reflect the rapidly changing face of biology; and many of the consequent alterations have been suggested by the student and teacher users of the texts. To all who have shown interest and aided us we express thankful appreciation.

WILLIAM D. MCELROY
CARL P. SWANSON

THAT A KNOWLEDGE OF CELLULAR STRUCTURE, FUNCTION, AND BEHAVIOR is crucial to an understanding of modern biology was evident as soon as the cell was clearly recognized as the basis of biological organization; but the interrelations of cell biology with biochemistry, physiology, and genetics remained obscure and tenuous until structure and behavior could be visualized within the same dimensional orders of magnitude. The electron microscope, with its great powers of resolution, has made these interrelations more secure and meaningful, and the convergence has led to an enormously enriched and coherent view of the cell in all aspects of division, inheritance, differentiation, development, aging, and disease.

In this volume I intend to deal with the structure, function, and behavior of normal cells and cell organelles, with examples from the animal, plant, and microbial kingdoms. (Biochemical and physiological topics, treated in other volumes of this series, have been minimized or generalized.) In this edition, in keeping with the continual expansion of cell biology, the number of illustrations—particularly of electron micrographs—has been increased; and the single chapter in the previous editions on cellular structure has been reorganized into three, dealing with membranes, particulates, and the nucleus. The remaining chapters have been extensively revised where new information or superior examples or illustrations would serve to convey a more adequate and appropriate view of the cell.

I hope, therefore, that this edition will continue to serve the student and the teacher as a sound introduction to cell biology, whether used by itself or in conjunction with other instructional materials. I am well aware of the debt that I owe to those who have helped to make this book what it is, and to all I am grateful indeed.

CARL P. SWANSON

THE CELL

CONTENTS

THE CELL

The struggle to know is one of the most exciting dramas of history, and every man who ever tried to learn anything has enacted it for himself to some extent.

RICHARD R. POWELL

1 THE CELLULAR BASIS OF LIFE

THE BIOLOGICAL SCIENCES OF TODAY CAN TEACH US A PROFOUND AND meaningful lesson whose validity we may no longer doubt. Their lesson, in brief, is that biology not only recognizes the uniqueness of every living organism, but also supplies compelling evidence for a rational explanation of this uniqueness in molecular terms. The purpose of this book is to explore the cell, which we now know to be the structural and functional basis of this uniqueness.

The details of nature are revealed to him who has the eyes to perceive, the patience to observe, and the ability to analyze. It is through these details that the beauty, diversity, and unity of nature are revealed. We have come, as a result of observations, to recognize that the universe around us, both living and nonliving, has structure, and that this structure exists because the elements within it possess organization. Common experience tells us that nature is made up of matter with certain describable and measurable characteristics. We realize that each of these "things" has a distinguishable uniqueness that we detect through touch, taste, hearing, smell, or sight. With our unaided senses, we have no difficulty in distinguishing the sky

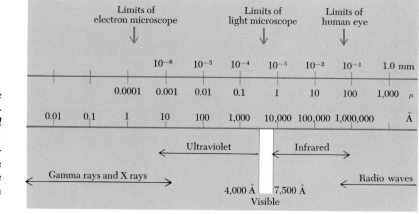

Figure 1.1 The electromagnetic spectrum on a logarithmic scale, measured in millimeters, microns, and angstrom units: 1 μ (micron) = 0.001 mm (millimeter) = 10,000 Å (Angstroms). The approximate lower limits of resolution of the human eye, the light microscope, and the electron microscope are given.

and the land from the water, a gas and a solid from a liquid, the living from the nonliving. On a more refined level, we can distinguish degrees of roughness, intensity and shade of color (if we are not color-blind), and an acid taste from one that is salty, sweet, or bitter. But human powers of sensory discrimination are limited. We all know, for example, that water, steam, and ice are made up of the molecule H_2O, and that they have different characteristics, but our ordinary senses cannot tell us why this is so. We hear within only a certain range of sound waves, and see only that portion of the light spectrum called the visible region (Figure 1.1). When we try to go beyond these limits, we can no longer directly comprehend the physical nature of things and must resort to instruments to penetrate areas outside our naturally circumscribed sphere.

Instruments, therefore, act as extended senses. Try to imagine, if you will, how much of your knowledge of yourself and the universe around you has been gained *only* through the use of your five senses as compared to that derived through the use of instruments. The 200-in. Hale telescope on Mount Palomar, for example, reaches across millions of light-years to bring distant galaxies of the macrocosm into view and to aid us in determining the age of the universe, its mode of origin, and its continuing evolution. Light microscopes and electron microscopes reach down into the microcosm to reveal other worlds, ordinarily invisible because of minute size. Similarly, photographic plates, more sensitive than our eyes, extend our use of light rays. Ordinarily, we can see only a minute portion of the electromagnetic spectrum (Figure 1.1), but by utilizing photosensitive surfaces we can detect the long infrared rays on one side of the spectrum and the short ultraviolet rays, X rays, and gamma rays on the other.

Whatever the means we use, we attempt to define the "things" we observe in terms of *units*, and the more refined our knowledge and the more powerful and discriminating our instruments and techniques, the more precise become our definitions of these units, that is, their limits, basic nature, and modes of aggregation into larger units. It would, indeed, be impossible for you to read these pages without understanding letters, the basic symbolic units of our language, or the numbers that make up our decimal or metric system. The periodic table of elements is another example of such a coherent system, and part of its great value lies in enabling us to predict what will happen under specified physical or chemical circumstances. One of the first goals of a science, therefore, whether it be physics, chemistry, or biology, is to determine the uniqueness of the units with which it is concerned, for unless such units are understood and accepted by everyone in a particular field, meaningful communication is difficult and scientific knowledge in that field cannot progress.

Science makes use of two kinds of units. Those used to describe or to measure time, weight, and distance are arbitrarily defined but we accept them universally as standards for the sake of convenience. Thus millimeters, microns, or angstroms can be used as convenient measuring units to define limits of the electromagnetic spectrum (Figure 1.1); kilometers, miles, or light-years are more useful for long distances. Such units as the electron, proton, and neutron, on the other hand, have a demonstrated physical reality that can be independently determined by anyone having the proper instruments and required knowledge.

It is the latter type of unit that we will investigate here, for the basic unit of life, the *cell*, is a physical entity. We can break up cells and extract selected parts for study much as the physicist breaks up atoms. We find that these cellular fragments can carry on many of their activities for a time; they may consume oxygen, ferment sugars, and even form new molecules. But these activities individually do not constitute life any more than the behavior of a subatomic particle is equivalent to the behavior of an intact atom. The disrupted cell is no longer capable of continuing life indefinitely; we therefore conclude that the cell is the most elementary unit that can sustain life, even though, as we shall see, the cell is highly complex. Viruses, on the other hand, are smaller and less complex than cells, but they cannot maintain life independently of the cells they parasitize (page 89).

Compared to the atom and the molecule, the cell is a unit of far greater size and complexity. It is a microcosm having a definite boundary, within which constant chemical activity proceeds. At ordinary temperatures, a chemically quiescent cell is dead. The cytologist (one who studies cells), therefore, seeks to identify the kinds of cells that exist, to understand their organization and structure in terms of their activities and functions, and to visualize the cell not only as a total entity (as, for example, the unicellular bacterium) but also as an integral part of the elaborate organs and organ systems of multicellular plants and animals.

The now familiar idea that the cell is the basic unit of life is known as the *cell doctrine*. It developed gradually through microscopical observations of the structure of many plants and animals, and eventually the presence of cells as a common structural feature of all biological organization was recognized. Two German scientists, M. J. Schleiden and Theodor Schwann, the former a botanist and the latter a zoologist, formally spelled out the cell doctrine in the years 1838 and 1839, but only after the cell and its contents had been generally identified by others as the improvement in microscopes and microscopical techniques of fixing and staining permitted the detection of finer details. Nevertheless the cell doctrine, however vague its beginnings and however long its developmental history, now ranks with Charles Darwin's *theory of evolution* and the *theory of the gene* as one of the foundation stones of modern biology.

The emergence of a great scientific generalization is generally a slow accumulative process; very few men and their ideas stand alone in the stream of time. The significance of the dates 1838 and 1839 and of the names Schleiden and Schwann, therefore, does not lie in a sudden discovery of cells for the first time: in fact, Robert Hooke, an Englishman, first saw the walls of cells in 1665 in a piece of cork as he was using his newly invented, primitive microscope (Figure 1.2). It was also Hooke who applied the word "cell" to designate the tiny structures he observed in the new world he had discovered.

Schleiden and Schwann were not the first to believe in, or advance, the idea that plants and animals are made of cells and cell products. During the seventeenth and eighteenth centuries many workers in Europe described cells and discussed their significance, and by 1800 good microscopes as well as the techniques for preparing cells for observation were becoming available. By 1800 there was a rather general acceptance of the idea that organisms are cellular, but there was much confusion over the definition of cells, the significance of their contents and of the cell walls, their mode of origin, and their

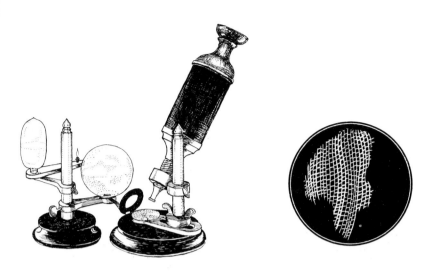

Figure 1.2 *Robert Hooke's microscope with which he observed the microscopic structure of cork and his drawing of it (in circle). Here, in his own words, is a description of his experiment: "I took a good clear piece of Cork and with a Pen-knife sharpen'd as keen as a razor, I cut a piece of it off, and thereby left the surface of it exceeding smooth, then examining it very diligently with a Microscope, me thought I could perceive it to appear a little porous; but I could not so plainly distinguish them as to be sure that they were pores. . . . I with the same sharp pen-knife cut off from the former smooth surface an exceeding thin piece of it, and placing it on a black object Plate. . . . and casting the light on it with a deep plano-convex Glass, I could exceedingly plainly perceive it to be all perforated and porous, much like a Honeycomb, but that the pores of it were not regular . . . these pores, or cells, were not very deep, but consisted of a great many little Boxes, separated out of one continued long pore by certain Diaphragms . . . Nor is this kind of texture peculiar to Cork only; for upon examination with my Microscope, I have found that the pith of an Elder, or almost any other Tree, the inner pulp or pith of the Cany hollow stalks of several other Vegetables: as of Fennel, Carrets, Daucus, Bur-docks, Teasels, Fearn . . . & c. have much such a kind of Schematisme, as I have lately shewn that of Cork."*

role in development. What Schleiden and Schwann did, however, was to take the loose threads of ideas and observations and weave them into a convincing doctrine that cells containing nuclei were the structural basis of organization for both plants and animals. Many of their ideas concerning cellular structure, function, and origin have since been proved erroneous, but by emphasizing the importance of the cell, they gave coherence to the biological thought of their time and focused attention on the one structure that had to be understood if biology was to advance beyond its purely descriptive stage. A study of the cells of such divers organisms as bacteria, orchids, and man aids in the understanding of the structure and function of all organisms, an impossibility prior to the acceptance of the cell doctrine. Who would have thought, prior to microscopical studies, that man and an orchid had anything in common?

Some 20 years after the announcements of Schleiden and Schwann, Rudolf Virchow, a great German physician, made another important generalization: *that cells come only from preexisting cells.* When biologists further recognized that sperm and ova are also cells that unite with each other in the process of fertilization, it gradually became clear that life from one generation to another is an uninterrupted succession of cells. Growth, development, inheritance, evolution, disease, aging, and death are, therefore, varied aspects of cellular behavior, even though each of these phenomena can also be viewed at higher and lower levels of biological organization.

Most generalizations have exceptions to them that cast doubt on their universal validity. This is true as well for the cell doctrine, but we shall defer a consideration of these exceptions until after we have examined cellular structure in some detail. Let us now consider what the cell doctrine, as presently interpreted, embodies in the way of solid ideas. These are essentially three in number:

In the first place, as we have already mentioned, the cell doctrine states that life exists only in cells; organisms are therefore made up of cells; the activity of an organism is dependent on the activities of cells, individually and collectively; and the cell is the basic unit through which matter and energy are acquired, converted, stored, and utilized, and in which biological information is stored and manipulated. We shall find, though, that we may have to modify this idea when we come to discuss the viruses, which lack a conventional cellular organization. Second, and as a direct corollary of the first generalization, the cell doctrine has embodied within it the idea that the continuity of life has a cellular basis, which is another way of stating Virchow's generalization. Now, however, we can be more explicit and enlarge upon this theme, adding that genetic continuity in a very exact sense includes not only the cell as a whole but also some of its smaller components, such as genes and chromosomes. The third idea is that there is a relation between structure and function. This has been called the *principle of complementarity;* it means, briefly, that the biochemical activities of cells occur within, and indeed are determined by, structures organized in a definite way. We shall encounter this idea again in our discussion of cellular components.

André Lwoff, the French microbiologist, has expressed the cell doctrine in yet another way: *

When the living world is considered at the cellular level, one discovers unity. Unity of plan: each cell possesses a nucleus imbedded in protoplasm. Unity of function: the metabolism is essentially the same in each cell. Unity of composition: the main macromolecules of all living beings are composed of the same small molecules. For, in order to build the immense diversity of living systems, nature has made use of a strictly limited number of building blocks. The problem of diversity of structures and functions, the problem of heredity, and the problem of diversification of species have been solved by the elegant use of a small number of building blocks organized into specific macromolecules. . . . Each macromolecule is endowed with a specific function. The machine is built for doing precisely what it does. We may admire it, but we should not lose our heads. If the living system did not perform its task, it would not exist. We have simply to learn how it performs its task.

Progress in the life sciences has not followed an even course, for it has been dependent on the development of ever more refined tools and techniques of analysis. This requirement has been especially true for cytology. Some cells may be large enough to see with the unaided eye. But to identify their internal organization we must magnify them greatly, and often use dyes that stain only selected parts of the cell.

Adequate magnification is as much of a problem for the cytologist as it is for the astronomer; the latter has to overcome great distances, the former very small sizes, in attempting to study in detail the objects they observe. For our purposes, the problem of magnification can best be considered in terms of *resolving power,* which is the ability of an optical system to reveal details of structure. In observing a double star (for example, one of those found in the handle of the Big Dipper, seen in the Northern Hemisphere), some of you will be able to discern but a single star; others, with better resolving power, will see two separate stars. In a compound microscope, the resolution of the first magnifying lens is the critical factor. As Figure 1.3 indicates, the lens nearest the specimen being examined, called the *objective* lens, is the key element of a compound microscope, because the projector lens, the *ocular,* can enlarge only what the objective has resolved.

The unaided human eye has a resolving power of about 0.1 mm

* André Lwoff, *Biological Order* (Cambridge, Mass.: M.I.T. Press, 1962), pp. 11, 13.

(millimeter). Lines closer together than this will be seen as a single line, and objects that have a diameter smaller than this range will be invisible or seen only as blurred images. The human eye, however, has only slight powers of magnification; each of us must calculate sizes mentally, and experience is probably the largest factor in our ability to judge accurately. Microscopes, of course, both resolve and magnify, but their resolution is limited by the kind of illumination used. Objects that are closer than one-half the wavelength of the illuminating light cannot be clearly distinguished in a light microscope. Thus, even with the most perfectly ground lenses, and with white light having an average wavelength of 5,500 Å (angstroms), an oil immersion objective cannot resolve two discrete points separated from each other by a distance less than 2,700 Å, or 0.27 μ (micron). Since many parts of cells have smaller dimensions, their presence and struc-

Figure 1.3 Schematic representation of the optical systems of the eye, the light microscope, and the electron microscope.

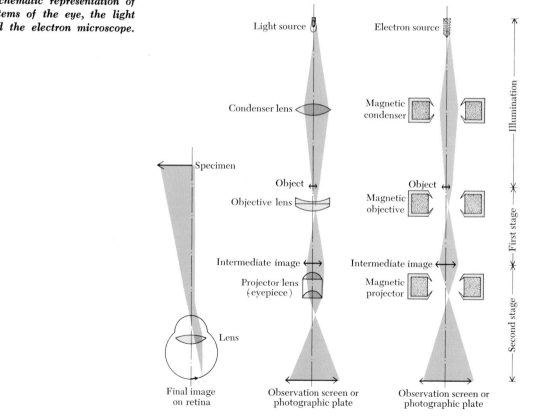

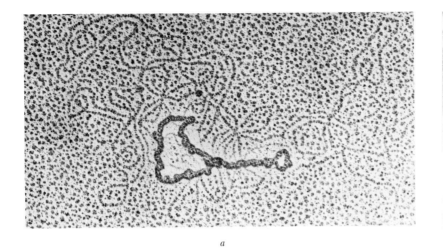

a

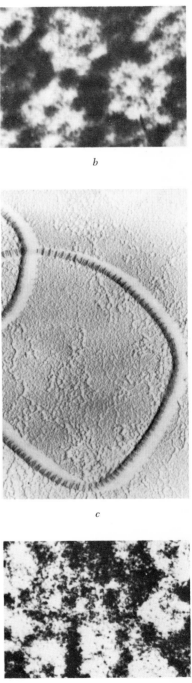

b

c

d

Figure 1.4 **Molecules of biological importance photographed in the electron microscope.** (a) Molecules of nucleic acid; the dark heavy strand is the hereditary material of a virus that has invaded a bacterium, Salmonella; the longer, lighter strand is the hereditary material of the bacterium; the background spots are of film emulsion. Protein molecules: (b) aggregates of the carbon dioxide (CO_2)-fixing enzymes extracted from spinach chloroplasts; (c) collagen, an animal protein found prominently in cartilage, showing its characteristic banded structure; (d) a calcium-dependent enzyme (ATP-ase) extracted from spinach chloroplasts. (Courtesy of Dr. E. Moudrianakis.)

ture were undetected until a means of greater resolution was found.

The *electron microscope* provides this increased resolving power by making use of "illumination" of a different sort. High-speed *electrons* are employed instead of light waves. As these pass through the specimens being viewed, parts of the cell absorb or scatter electrons differentially, thus forming, by means of lenses, an image of the specimen on an electron-sensitive photographic plate or fluorescent screen. The human eye is not stimulated by electrons, hence the need for plates or screens. The "optical" system is similar to that in the light microscope (Figure 1.3) except that the "illumination" is focused by magnetic lenses instead of conventional glass lenses.

When electrons are accelerated through the microscope by a potential difference of 50,000 volts, they have a wavelength of about 0.05 Å. This is 1×10^{-5} that of average white light. An electron microscope can thus theoretically resolve objects that are separated by a distance of one-half of 0.05 Å, or 0.025 Å. This dimension is far

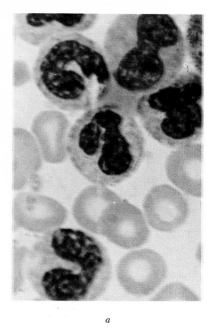

a

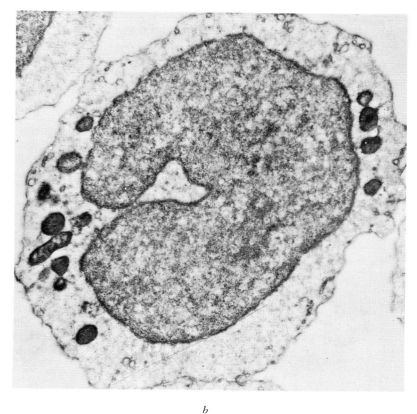

b

*Figure 1.5   Resolution differences.
(a) A photomicrograph, taken through
a light microscope, of a group of human lymphocytes. (b) Electron micrograph of a comparable cell.*

less than the diameter of an atom (the hydrogen atom has a diameter
of 1.06 Å), but owing to limitations in design, the actual resolving
power of the best modern instrument is about 2 to 3 Å. At this level,
individual atoms cannot be destinguished, but large molecules of biological importance are readily visible (Figure 1.4). In approximate
figures, then, the human eye can resolve down to 100 μ, the light
microscope to 0.2 μ, and the electron microscope to 0.001 μ. Or, to
put it another way, if the human eye has a resolving power of 1, then
that of the light microscope is 500, and that of the electron microscope 100,000. The electron microscope has thus opened up a whole
new domain to the cytologist, by making a number of cellular structures visible.

The increasing degree of resolution made possible by advancements
in microscopy is indicated in Figure 1.5. How much more refined
we can become in our visualization of cellular structure depends upon

improvement in lens construction for the electron microscope, and on the reliable preservation of the fine structures of the cell. An electron microscope that would resolve structure at the level of 1 Å would enormously expand our field of vision, for it would enable us to visualize molecular organization as well as molecular aggregates.

In the study of cells more than just clear resolution and high magnification are required; the parts of the cell must be clearly distinguished from their immediate surroundings. In the electron microscope, this contrast is possible because some structures are or can be made more electron "dense" than others, and photosensitive film is darkened to the degree that it is struck by the electrons passing through a specimen. Since the degree to which electrons are scattered is a function of the mass of the atom, and since the light atoms of organic materials—hydrogen, carbon, nitrogen, and oxygen—have little scattering power, the parts of the cell to be examined may be selectively "stained" with heavy metals in order to show contrast. Heavy metals such as osmium, bismuth, uranium, and manganese are customarily used. Contrast is equally difficult to achieve with the light microscope because most parts of the cell are transparent to light. To overcome this problem, the cytologist uses the proper killing agent (fixative) and stain to color the parts he wishes to examine. Literally hundreds of fixing and staining procedures are known; they are the cytologist's recipes, and he continually improves upon them in his search for better ways to study cells. Since many molecules, because of their chemical makeup, will selectively absorb particular dyes, the staining procedures are used not only to reveal cellular structure but also to assist in the identification and distribution of cell molecules that could not be detected in any other way.

Parts of cells can also be selectively studied through the use of molecules containing radioactive atoms, particularly phosphorus 32 (P^{32}), carbon 14 (C^{14}) and tritium (H^3), an isotope of hydrogen. When radioactive molecules are taken up and attached to particular parts of the cell, their presence can be detected through *autoradiography*. That is, a thin layer of photographic emulsion is spread over the flattened cells, and as the radioactive atoms disintegrate, the rays or particles released from these atoms pass into the emulsion and cause a darkening of the emulsion much in the manner that exposure to light causes a photographic film to darken. When such cells are viewed under the microscope they may appear as in Figure 1.6a. The very active fields of *cytochemistry* and *histochemistry*, which deal respectively with the chemistry of cells and tissues, are greatly dependent on these procedures.

A living cell, however, is always more fascinating than a dead one.

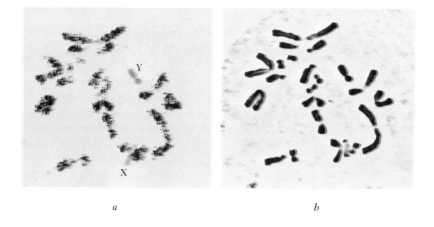

a   b

Figure 1.6 **A metaphase cell of the Chinese hamster,** *photographed in the light microscope (b), and by autographic methods (a). The black dots in illustration (a) are formed when the radioactive tritium (H³) atoms in the chromosomes disintegrate and cause a blackening of the emulsion covering the cell.* (Courtesy of T. C. Hsu.)

To watch cells divide is to witness one of the most dramatic of biological phenomena. The *phase-contrast microscope* permits the cytologist to do this. Various parts of a living cell stand out in sharp contrast when light, passing through these parts, is thrown out of phase, hence, the name of the instrument. Figure 1.7 is a photograph of a living human cancer cell, taken through a phase-contrast microscope. Under the conventional light microscope, such a cell would appear almost structureless.

Another way to study living cells is to "grind" them up and examine their parts. This is done with a special mortar and pestle to burst the cells and release their contents into a solution. When this

Figure 1.7 **A human cancer cell (HeLa strain),** *grown in tissue culture and photographed through a phase-contrast microscope. The nucleus with its nucleolus is seen in the center; the white mass is liquid refractile material taken in from the culture medium; the slender rods inside the cell are mitochondria; and the outer fine projections are microfibrils formed by free cells when cultured in test tubes.* (Courtesy of Dr. George O. Gey.)

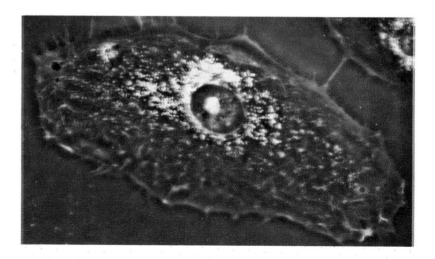

solution is centrifuged at carefully regulated speeds, the denser portions settle out at lower speeds, the less dense ones at higher speeds. Once separated, each portion can be analyzed for chemical structure or content, or tested for chemical activity, since in the test tube some parts continue to function for a while as they did in the intact cells.

Living cells of plants, animals, and humans can also be cultured with ease in much the same way as are bacteria or fungi, and subjected to a wide variety of procedures in order to test their response and future behavior. These cells can be probed by microneedles, dissected by the techniques of microsurgery, and injected with solutions through micropipettes. The cytologist, therefore, has a large and varied arsenal of instruments and a whole battery of procedures for making the cell give up its secrets. Any one tool or technique, however, is usually not enough, and several methods are often employed before an answer can be found to the particular problem being investigated. Such knowledge as we have already garnered from the cell has strengthened our belief that it is the basis of life, and at the same time has made us acutely aware of how little we know of its many complexities.

2 THE STRUCTURE OF CELLS: MEMBRANE SYSTEMS

EVERY BIOLOGICAL ACTIVITY INVOLVES ONE OR MORE CHEMICAL RE-
actions. Breathing, walking, seeing, tasting, thinking, or even just existing requires energy. This energy is derived from chemical reactions
that take place within individual cells. Furthermore, these energy-
requiring activities make little sense unless we also consider the structures related to them: lungs and diaphragm in breathing; muscles and
bones in walking; the lens, retina, and optic nerve of the eye in seeing. These more obvious structures are made up of cells or cell products, and their internal molecular arrangement determines their activity
and appearance.

We can therefore consider the cell to be an organized and directed
chemical factory, with matter taken in and transformed, and energy
acquired, converted, stored, and/or utilized. It may, of course, be a
general-purpose factory, capable of performing all the services and
of manufacturing all the products necessary to continue life; this must
obviously be true in unicellular organisms. Or it may be a specialty
shop, doing only a single kind of job—for example, nerve cells for
communication or muscle cells for movement. Regardless of its nature,

however, a cell, like a factory, must possess an organization in order to be efficient; it must contain a controlling center and pathways of communication that somehow tell it what to do and when to do it, a source of supplies, a source of energy, and the machinery for making its product or performing its service. It is not surprising, then, that cells, despite their great variety of shapes, sizes, and functions, share many common features. If a cell becomes specialized, we might expect to find a change in organization and, possibly, the appearance of new parts but not at the sacrifice of basic features. For this reason, the biologist considers that *form and function* are inseparable phenomena; to put it another way, organized activity is associated with an organized arrangement of parts.

That biologists recognize the uniqueness of every living organism should alert us to the fact that there is no "typical" plant or animal that is representative of all plants or animals. We can extend this concept to the level of the cell, for each cell, or cell type, is equally unique. For purposes of discussion, however, we can think of a "typical" cell that contains the features we wish to examine. This cell is bounded at its outer living limits by a *cell,* or *plasma, membrane,* which separates it from the environment and through which materials entering or leaving the cell must pass. Our typical cell contains a *nucleus.* This is the center of control of cellular activities; even though the nucleus is necessary for the continued existence of the cell, a cell without a nucleus may continue to function for varying periods of time. But a cell without a nucleus is a cell without a future, because it cannot give rise to viable progeny. The remainder of the cell is the *cytoplasm.* Here are contained the various membrane systems, particulate organelles, and soluble components that, through chemical reactions, control the problems of synthesis and of energy transfer and conversion necessary for the tasks that cells perform. We still lack many of the pieces of information needed to understand fully the coordinated behavior of this cellular factory, but the major features are reasonably clear.

Much of our understanding has come through the integration of discoveries by biochemists and electron microscopists, the former providing knowledge of the chemical pathways in cells and the latter, new concepts relating to organized cellular morphology. It is now possible to speak with some assurance about the *fine structure* of the cell in molecular terms and to view the various structures in the light of their function.

Here it should be mentioned that the use of the electron microscope in revealing the fine structure of cells was made possible only by the parallel development of techniques for the cutting of ultrathin slices of cells, or of subcellular fractions which have been aggregated by centrifugation. An ordinary microtome, used in connection with light microscopy, can cut sections about 4 to 5 μ thick. Thus, a cell 30 μ in diameter could be sliced into six or seven pieces. Visible light is not seriously absorbed in these slices except where stain was deposited. However, such sections would be too thick for use in electron microscopy, since this amount of material would be opaque to an electron beam, and little or nothing of cellular morphology would be revealed. By new sectioning techniques, slices between 100 to 500 Å (0.01 to 0.05 μ) in thickness can now be cut. A comparable cell could thus yield about 600 cross sections. This refinement means, of course, that the electron microscope, because of its tremendous powers of resolution, can reveal only very minute portions of a cell at any given time; in a sense, breadth of vision is sacrificed for intimate details. A comparison of various photographs in this book taken through the light and electron microscopes will convey something of the great steps forward that have been achieved in resolution during the past 20 years.

In order to preserve cells for sectioning, new techniques of fixation and embedding also had to be developed. The most successful fixatives are osmium tetroxide, potassium permanganate, and glutaraldehyde, and thin sectioning requires that the material be embedded in plastic resins instead of the usual paraffin. Uranyl acetate and lead citrate as well as osmium and manganese compounds provide the contrasts needed for clear resolution.

THE CELL SURFACE

Before the electron microscope came into general use, it was assumed that all cells had an outer limiting membrane that separated each one from the environment and thereby preserved the identity of the cell as a basic unit of organization. This inference was based on the fact that cells could swell or shrink, and that when their surface was torn by a dissecting needle the contents would ooze out. Studies with the electron microscope have confirmed the universal existence of this plasma membrane and emphasized its fundamental importance for cellular integrity. They have also shown that it is invisible in the light microscope because it is about 100 Å (or 0.01 μ) thick, and thus below the limits of ordinary resolution.

The trilaminar nature and the possible molecular makeup of the plasma membrane are depicted in Figure 2.1. The term "unit membrane" has been applied to this structure, although it should be recognized that membranes from various cells and organisms vary in

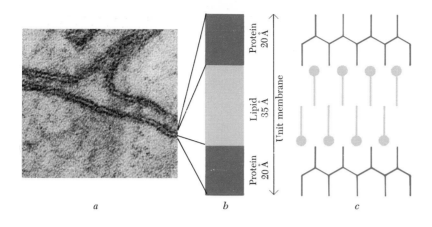

Figure 2.1 *(a)* *An electron micrograph of cell membranes at the juncture of three cells.* *(b)* *Diagrammatic representation of the structure of a single unit membrane, with its outer protein layers, each 20 Å thick, and an inner lipid layer about 35 Å thick.* *(c)* *The double nature of the lipid portion of the membrane (tadpole shapes) and the two outer protein layers.* (Photograph courtesy of Dr. J. D. Robertson.)

thickness, molecular composition and function. The idea that the membrane has a lipoprotein (fat plus protein) composition preceded studies with the electron microscope, and arose out of work done primarily on the permeability and surface tension of various kinds of cells, as well as through the analysis of red blood cell ghosts (the cell membrane free of the inner contents of the cell). Furthermore, it is possible to mix phospholipids and protein in water, and to show that they will assemble into membranes in the same manner as that found in normal cells. The protein constituent of the membrane gives the cell structural integrity and flexibility. Since protein molecules, which are long and complex, can fold or unfold, the membrane can expand or contract, thus providing through molecular spacing a possible means of control over which molecules can enter the cell from the outside environment, or pass to the environment from inside the cell. Such a membrane is said to be selectively permeable, with the degree of permeability related not only to the nature of the penetrating molecule but also to the state of the membrane at any given time. Such a membrane would also permit growth and movement, either for the cell as a whole or for localized regions, as is the case in an ameba when it moves along the surface of a glass dish. The mechanics of membrane contraction, expansion, or growth are not, however, fully understood.

The lipid portion of the membrane is suggested by the fact that fat solvents readily penetrate cells from the surrounding environment, while polar molecules such as water itself are hindered in their passage in and out of the cell. The idea that a double lipid layer is sandwiched between the two protein layers is derived from the observation that the lipid extracted from a known number of cells, and

THE STRUCTURE OF CELLS:
MEMBRANE SYSTEMS

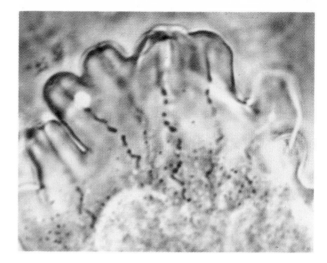

Figure 2.2 **Photograph of the edge of a living ameba,** *showing the pinocytotic channels (the dark lines converging toward the center of the cell). Liquids flow into the cell through these channels, to be pinched off as membrane-enclosed droplets; these eventually dissolve in the interior of the cell. (Courtesy of Dr. David Prescott.)*

spread as a monolayer on water, covered twice the area calculated for the surface area of the cells. Physiological, analytical, and microscopical studies are, therefore, consistent with each other as to membrane structure.

The basic structural pattern of unit membranes, as revealed by electron microscopy, does not imply a uniformity of molecular composition. This is indicated both by the variations in permeability exhibited by different cells, and by observed differences in thickness of the three layers. Furthermore, a variety of enzyme systems are associated with, or are actually an integral part of, the cell membrane, and these, together with the nature of the membrane, govern the rate of entry of molecules as well as the kinds of molecules which can enter or leave a cell. The source of variation in molecular composition, providing the possibilities of a mosaic pattern over the surface of a single cell, resides in the great variety of proteins and lipids that could enter into the structure of such membranes. Such mosaicism would provide for physiological differences in permeability and surface reactions, but would remain undetected structurally with present levels of resolution.

Other problems of membrane permeability are discussed in another volume of this series,* but it is of interest to us here that free cells, such as those in tissue culture, can take in materials from the

* W. D. McElroy, *Cell Physiology and Biochemistry* (2nd ed.) (Englewood Cliffs, N.J.: Prentice-Hall, Inc., 1964).

liquid environment by one or both of two additional processes, *pinocytosis* and *phagocytosis*. The former name is derived from the Greek words for "drink" and "cell," and the process is literally a drinking phenomenon. The flexible plasma membrane forms a channel to get liquids into the cell, and then pinches off pockets that are incorporated into the cytoplasm to be digested (Figure 2.2). By this device, large molecules and various ions incapable of passing through the membrane can be taken up by cells. In phagocytosis (Gr. *phagein,* to eat), arms of cytoplasm engulf droplets of liquid containing solid material, such as bacteria, and draw these materials into the cytoplasm, where the digestive enzymes break the engulfed material down into usable fragments (Figure 2.3).

We can, therefore, consider the plasma membrane to be a portion of the living cell. This point of view is reinforced by the intimate connections of the plasma membrane with other internal membrane systems (Figure 2.4), its capability of limited repair if torn or punctured by a needle, and its activity in a cell exhibiting pinocytosis, phagocytosis, or mobility. We must recognize, however, that the membrane is elastic, changeable, and pliable in some cells, quite rigid and unyielding in others; thin in a cell such as that depicted in Figure 1.7, thick in some marine invertebrate eggs; smooth in an ameba, ciliated in a paramecium. We must recognize further that the cell can secrete products that, adhering to the cell surface, can determine the physiological properties of the cell. These surface substances are generally some form of polysaccharide (a long-chain sugar derivative) and would include such structures as the heavy cell walls of plants, the cuticle covering the surface of insects, and the thick surface of many eggs and bacterial cells. For example, red blood cells of the A, B, and O types are indistinguishable morphologically but are readily separable on the basis of their clumping reaction—a cell-surface phenomenon—when exposed to blood sera of different kinds. Cells from the same tissue also exhibit the ability to "recognize" one another when in liquid suspension. For example, if heart and kidney cells from a chick embryo are dissociated to form a mixed single-cell suspension, and then allowed to remain undisturbed in culture, heart cells seek out and aggregate with heart cells, kidney cells with kidney

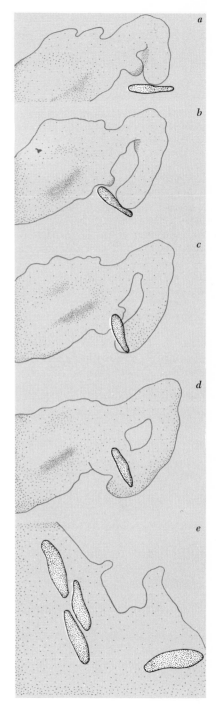

Figure 2.3 **Phagocytosis as observed in an ameba.** *The arm of cytoplasm, coming in contact with a paramecium, surrounds it and then draws it into the cytoplasm where it can be digested. (a–d) The process of enveloping. (e) A portion of the ameba containing several phagocytized paramecia.*

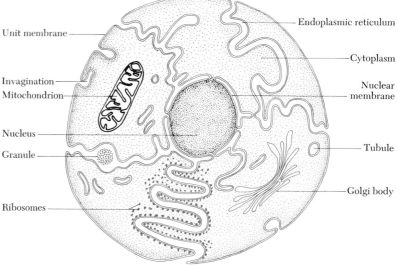

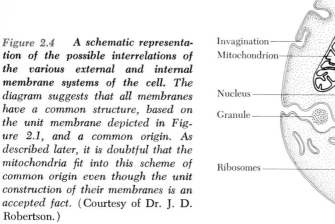

Figure 2.4   A schematic representation of the possible interrelations of the various external and internal membrane systems of the cell. The diagram suggests that all membranes have a common structure, based on the unit membrane depicted in Figure 2.1, and a common origin. As described later, it is doubtful that the mitochondria fit into this scheme of common origin even though the unit construction of their membranes is an accepted fact. (Courtesy of Dr. J. D. Robertson.)

cells. The reaggregation is a surface phenomenon, and can be prevented by an alteration of the membrane surfaces. Cell membranes, therefore, possess discrete properties. Some of these properties may be conferred upon the plasma membrane by associated surface molecules—for example, the glyco- or mucoproteins of red blood cells—but the membrane itself may possess equally distinct arrangements of parts that determine its ultimate functional character.

We should not leave the impression that this structure is a simple envelope, comparable to a thin plastic bag surrounding the cell contents. Three examples will show how its contours may be modified in certain specialized cells. Figure 2.5 is of a columnar epithelial cell in the lining of the small intestine of the mouse; such a cell is active in the absorption of digested food. The plasma membrane here is greatly convoluted, and contains regions that show intimate connections—*desmosomes* (Figure 2.6)—with adjacent cells, while the upper surface of the cell forms a "brush border" in which the many projections—*microvilli* (Figure 2.7)—provide a tremendous absorption area. Each such cell would have upwards of 2,000 to 3,000 individual microvilli, while a square millimeter of the intestine would have as many as 200,000,000. In a gross way, the lining of an intestine and the convoluted surface of each cell is comparable to the absorptive surface of a bath towel.

The relation between a nerve cell and its associated *Schwann,* or

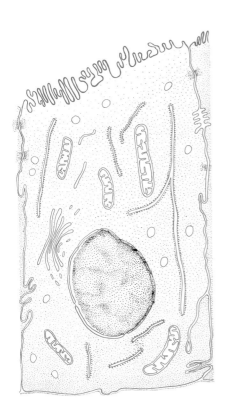

Figure 2.5 *Schematic drawing of a columnar epithelial cell from the small intestine of a mouse, showing how the plasma membrane can be folded along its sides, and greatly convoluted (top) to form the microvilli that are active in absorption (see Figures 2.6 and 2.7 for details).* (H. Zetterqvist, doctoral thesis, University of Stockholm, 1956.)

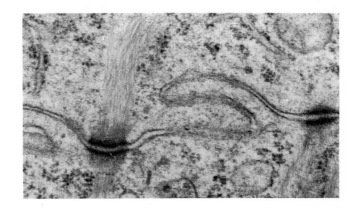

Figure 2.6 *The convoluted membranes of two adjacent columnar epithelial cells, showing two prominent desmosomes. Microtubules extend from the desmosomes into the cytoplasm, and clusters of ribosomes are also evident.* (Courtesy of Dr. C. Philpott.)

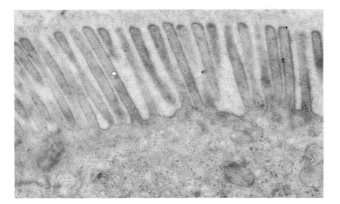

Figure 2.7 *Microvilli projecting from the surface of a columnar epithelial cell. In the light microscope the villi would appear as the "brush border."* (E. Leitz, Inc.)

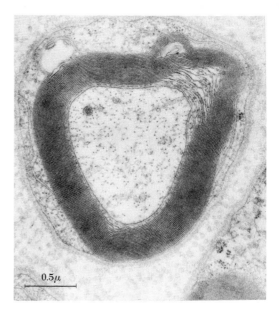

Figure 2.8 **Electron-micrograph cross section through a nerve fiber, with the membranes of the Schwann cell wrapped around the central axon. The cytoplasm of the Schwann cell can be seen outside the membranes and between the disrupted layers of membranes at the upper right. The basement membrane (see Chapter 3) of extracellular materials lies on the extreme outside, surrounding the entire structure.** (Courtesy of Dr. L. G. Elfvin.)

0.5μ

satellite, cells presents another example of membrane flexibility. Here the entire plasma membrane of one cell forms an elaborate structural system around a portion of another cell. Figure 2.8 depicts a nerve fiber, or *axon*—the extended process of a nerve—with the associated Schwann cell wrapped around it. The development of the spiral pro-

Figure 2.9 **Schematic representation of the progressive envelopment of an axon by the membranes of the Schwann cell, as described by Dr. Betty B. Geren. Such an axon is said to be myelinated.**

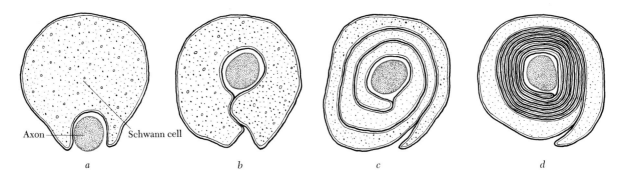

Axon — Schwann cell

a   *b*   *c*   *d*

Figure 2.10 *Schematic representation of a rod (light-receptor) cell in the retina of the guinea pig. The discs at the top of the cell are folded and refolded to provide many layers of membranes, each of which contains light-sensitive pigments on its surface. The mitochondria are concentrated just below the light-sensitive area; the rod nucleus is also identified. At the base, each cell has an intimate connection with a nerve fiber.* [F. S. Sjöstrand, *International Review of Cytology*, 5 (1956). New York: Academic Press, Inc.]

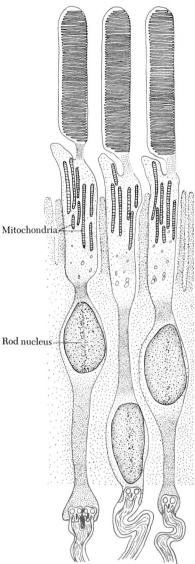

ceeds as in Figure 2.9. The cytoplasm of the Schwann cell is largely squeezed to the outside, leaving the axon surrounded by a multilayered membranous system and isolated from its surroundings. The *myelin sheath*, as this layered structure is called, is thought to assist in the transmission of nerve impulses. The clarity of the trilaminar nature of each membrane tends to be lost through the fusion of adjacent membranes.

A third modification of the cell membrane can be seen in the light-receptor cells (rods and cones) of the vertebrate eye (Figure 2.10). The outer segment of these cells is made up of a series of flattened discs, from 500 to a 1,000 in some cells, stacked one on top of the other like coins. The discs are derived from foldings of the cell membrane, but they tend to break free from the membrane and appear as free-floating structures, at least in sections prepared for electron microscopy (Figure 2.11). This significance of these discs is that they represent the light-receptor surfaces of the eye. It has been

Figure 2.11 *A small portion of a retinal cell, with the folded membranes where the light-sensitive pigment may be layered. That the membranes originate from the cell membrane is not shown here.* (Courtesy of Dr. J. D. Robertson.)

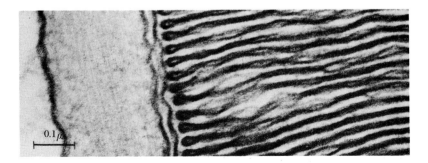

estimated that a single rod possesses 30 to 40 million receptor molecules, and, in some manner not yet determined, these are layered on the discs. The light-receptive surface area of a rod or cone is, therefore, enormous. If the diameter of the outer segment is 30 μ, the surface area of a disc is 942 μ^2 if the pigment is on one side only, 1,884 μ^2 if on both sides. 500 discs would give a surface area per cell of 471,000 μ^2 if single-layered with pigment, 942,000 μ^2 if double-layered.

THE ENDOPLASMIC RETICULUM

If a living cell has its plasma membrane ruptured, the cytoplasm oozes out as a slightly viscous liquid. This suggests that there is little structural organization of cytoplasmic materials, but this view of the cytoplasm has been drastically altered by the discoveries of the electron microscopists. In cells of almost every kind there exists an elaborate membrane system that is now recognized as an important part of the cell for the manufacture of cellular products. This is the *endoplasmic reticulum* (ER), or *ergastoplasm* (Figure 2.12), a membrane-limited cisternal (saclike) system extending, to varying degrees, from the *nuclear membrane* (see Chapter 4) on the inside to the plasma membrane on the outside of the cell. The outer nuclear membrane is often

Figure 2.12 **The endoplasmic reticulum (ER) in parotid (salivary-gland) acinous cells of the mouse. (a) The ER above the mitochondrion is of the rough, or granular, variety, containing ribosomes, and is much more highly organized than in the area immediately below; many of the membranes end blindly in the cytoplasm. (b) The mass of rough ER membranes is a continuously branching and interconnected system, which is also connected with the outer portion of the nuclear membrane (arrow).** (Courtesy of Dr. H. F. Parks.)

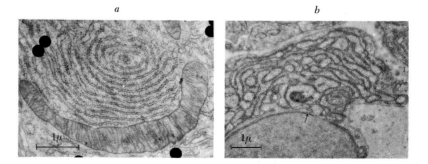

THE CELL

24

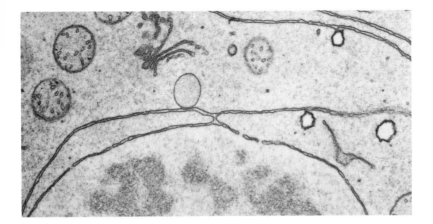

Figure 2.13   Electron micrograph of a portion of a roottip cell of a plant, showing the relation of the ER and the nuclear membrane. (Courtesy of Dr. G. Whaley.)

continuous with the ER, as depicted in Figure 2.4 and as seen in Figures 2.12 and 2.13. The membranes of the ER have the trilaminar structure of the cell membrane. The relation of the ER to the nuclear membrane is also suggested by the fact that when a cell divides, the new nuclear membranes are formed in part from nearby remnants of the ER.

The ER is of a variable morphology, and can be present in varying amounts. Each kind of cell, and its state of metabolic activity, can, in fact, be characterized by the appearance of its ER. It is, furthermore, a labile cytoplasmic system since it can alter its nature rather rapidly. Figure 2.12 represents two states of the ER in a parotid cell from the salivary gland of a mouse. In Figure 2.12(a), the sac-like ER is much flattened, like collapsed balloons, with the individual cisternae arranged in parallel, semicircular aggregates. In Figure 2.12(b), the cisternae are more expanded, and appear to form a continuously branching and interconnected system. Some cells may have substantial amounts of ER localized in given areas while being free of them in other areas.

The membranes of the ER are of two kinds: rough or smooth. Both may be present in the same cell, for example, in mammalian liver cells, but the presence of either is an indication of the particular synthetic capacity of the cell. The rough, or *granular*, ER is found in great abundance in cells actively engaged in protein synthesis, but the synthesizing capacity is not solely a property of the membrane itself, but rather is a complex process involving cooperative interplay of nucleus, ER and *ribosomes*. The ribosomes are attached to the membrane, and produce the rough appearance (Figure 2.14). The

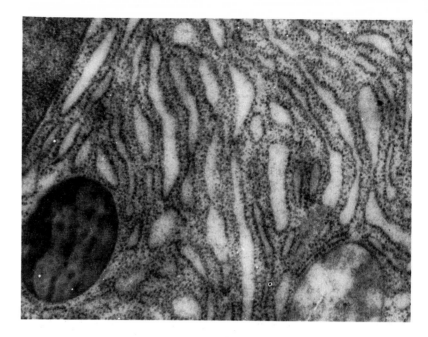

Figure 2.14   Electron micrograph of a portion of a pancreatic acinous cell, showing the rough ER. The ER forms flattened saclike arrangements, with the ribosomes on the outside of the sacs, or cisternae; the ribosomes do not project into the sacs, which are the clear areas. (Courtesy of Dr. G. Palade.)

ribosomes are electron-dense and rich in ribose nucleic acid (RNA) and protein, are about 250 Å in diameter, but consist of two separable parts which are distinguishable from each other by their rates of sedimentation when centrifuged (see page 13). Their role in protein synthesis will be discussed later, but since they can also be free in the cytoplasm without losing their synthetic capability, the role of the ER appears to be that of providing a system of communication in the cell which permits segregation of the products formed, and channels for the transportation of products either to other parts of the cell or to the outside.

Whether there are enzymes which, as proteins, also function as an integral structural part of the granular ER is open to question. It is suggested by the fact that there is no sharp morphological discontinuity between the granular and the *agranular,* or smooth, ER. The latter are free of attached ribosomes, hence their designation as agranular, and certain enzymes appear to be a part of the membrane since they cannot physically be separated from the membrane fragments. Despite physical continuities and overlapping morphological characteristics, the two kinds of ER are generally distinct in appearance, the smooth ER being in the form of tubules (Figure 2.15) rather than flattened cisternae. It is prevalent in a number of different kinds of

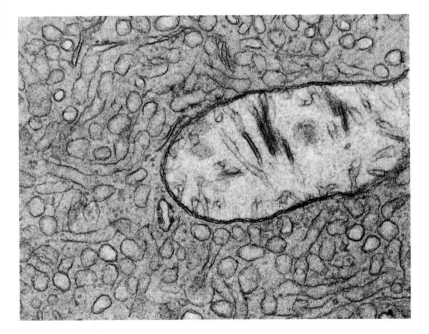

Figure 2.15 **Electron micrograph of a portion of a cell from the testis of an opossum, showing the smooth, or agranular, form of ER. (Courtesy of Dr. D. Fawcett.)**

cells, and appears to perform a variety of functions. In the liver it is concerned with lipid and cholesterol metabolism, and may be markedly increased in amount by the administration of lipid-soluble drugs such as phenobarbital. In cells of the testes and adrenal gland, the smooth ER is concerned with the synthesis of steroid hormones, while it is involved in the secretion of chloride ions in certain cells lining the mammalian stomach and the gills of fishes. The membranes of the smooth ER are generally thinner than other unit membranes, but their character is similar in cells of quite diverse function.

Named after its discoverer, the *Golgi complex* (it is also known as the *dictyosome*) is another system of unit membranes found in a wide variety of both plant and animal cells. It is comparable to the agranular ER in that its membranes lack attached ribosomes, but although it may show continuity with the ER it is more often discontinuous, readily stained by osmium or silver-containing dyes, and more compact in morphology. As Figure 2.16 indicates, it consists of a parallel array of flattened cisternae arranged in straight or curved bundles, with numerous small vesicles, 400 to 800 Å in diameter, clustered at the

THE GOLGI COMPLEX

THE STRUCTURE OF CELLS:
MEMBRANE SYSTEMS

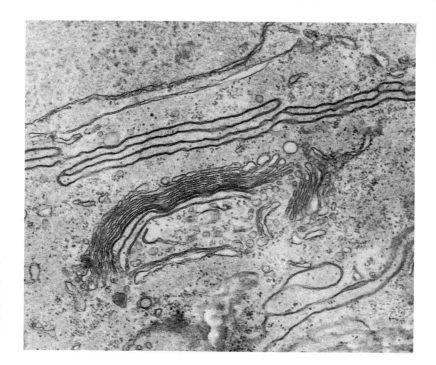

Figure 2.16   Electron micrograph of a Golgi complex in an animal cell. The parallel flattened sacs are usually curved slightly, and they tend to bud off small vesicles at the ends of the sacs. A Golgi complex in a plant cell can be seen in Figure 2.13.

ends of the cisternae. The vesicles originate from the cisternae by being pinched off, and frequently contain secretion products of both soluble and particulate nature.

The function of the Golgi complex is not yet fully understood. In cells known to produce and secrete carbohydrates, it seems likely that at least part of their synthesis is carried on by enzymes of the Golgi complex, and that these carbohydrates are complexed in the Golgi region with protein produced elsewhere in the cell. In other cells whose primary role is the production of proteins, the Golgi complex appears to be an assembly area. The proteins produced on the ribosomes are channeled by the granular ER to the Golgi complex where they are concentrated and packaged for distribution outside of the cell. It is also apparent that this structure plays a role in lipid metabolism. Fats tend to accumulate in the cisternae, and the appearance of the Golgi complex in animal cells can be greatly modified by fasting or by changes in fat diet.

The Golgi complex also plays a role in the transformation of the spermatid into a mature spermatozoan (Figure 8.14). This aspect, however, is discussed later (page 117).

IN ADDITION TO THE MEMBRANE SYSTEMS FOUND IN MOST CELLS, THERE are a number of particulate structures also present, each of which performs a given function and has a structure consistent with that function. The most prominent of these particulates is the *nucleus,* within which the *nucleolus* can generally be seen. In the cytoplasm are found the *mitochondria* (singular, *mitochondrion*), *lysosomes, centrioles, vacuoles, ribosomes,* granules and tubules of various kinds, and, in plant cells, *plastids* which assume various forms and engage in various activities. A discussion of the nucleus and ribosomes will be deferred to Chapter 4, where a consideration will be given to their role in the control of cellular metabolism.

Mitochondria are found in every type of cell except those of bacteria and blue-green algae. They are also absent from the mature red blood cell, although they were present in the immature erythrocyte. In Figure 1.7 they can be seen as long, slender rods, but their shape and

MITOCHONDRIA

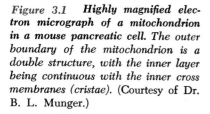

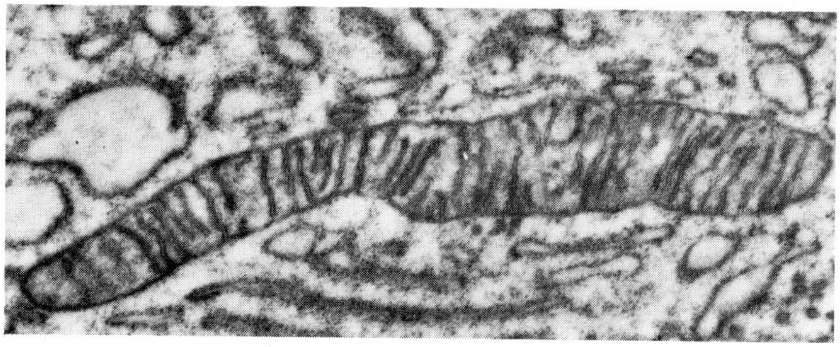

Figure 3.1 Highly magnified electron micrograph of a mitochondrion in a mouse pancreatic cell. The outer boundary of the mitochondrion is a double structure, with the inner layer being continuous with the inner cross membranes (cristae). (Courtesy of Dr. B. L. Munger.)

number are characteristic of the cell in which they are found. In the living cell, seen in tissue culture, they appear to be in constant motion.

Mitochondria range in size from 0.2 to 7.0 μ in diameter, and in form from spheres to rods to branching rods. The unicellular green alga, *Microsterias,* has a single mitochondrion per cell, whereas the giant ameba, *Chaos chaos,* may have up to 500,000. A mammalian liver cell, 25 μ in diameter, may possess as many as 1,000, a kidney cell about 300, and a sperm cell as few as 25. Plant cells appear to have relatively few when compared to an animal cell of similar size, but where cellular activity is high they tend to cluster. Such aggregation is not accidental, for as is now realized, these organelles play an important role in cellular economy.

Let us first consider their structure, which remained unknown until revealed through electron microscopy (Figure 3.1). Each mitochondrion is bound on its outside by a smooth unit membrane about 60 to 70 Å in thickness. Separated from it on the inside by a clear space of about 60 to 80 Å is an inner membrane, also of unit construction, which is variously folded into thin *cristae* extending into the center, or *matrix,* of the mitochondrion. The cristae may extend from wall to wall, or only part way across the matrix; they may be relatively few in number, leaving the majority of the mitochondrion occupied by matrical substance, or they may be so numerous as to nearly

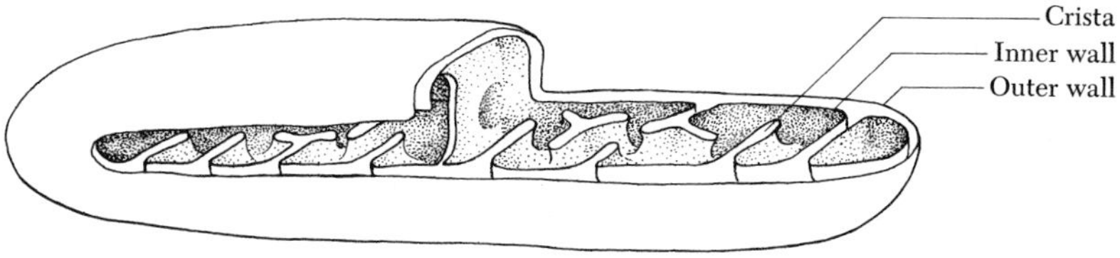

Figure 3.2 Schematic drawing of a typical mitochondrion. (Courtesy of Dr. A. H. Lehninger.)

Crista
Inner wall
Outer wall

clog the matrix with membranes. Figure 3.2 shows a schematic drawing of a liver mitochondrion, which is representative of the arrangement of cristae seen in mammalian cells. Occasionally the cristae may be in the form of tubules rather than septa, as, for example, in *Paramecium* and many plant cells, but whatever their form, they provide a greatly increased surface area on which sequential chemical reactions can take place. Enzyme complexes associated with the cristae form an inseparable and integral part of the structure of the membranes. There is a correlation, therefore, between the number of cristae per mitochondrion and chemical activity going on in the mitochondrion.

Figure 3.1 is a "positive" image of a mitochondrion in longitudinal section, that is, the electron-dense stains—osmium or potassium permanganate—are incorporated into, or are captured by, the membrane itself, giving the unit structure depicted in Figure 3.2. Negative staining with phosphotungstic acid reveals added details of structure in the cristae. The membrane itself does not stain, but rather is highlighted by an electron-dense background (Figure 3.3), revealing that the inner surfaces of the cristae are studded with small particles about 70 to 90 Å in diameter that are attached by short stalks. The arrangement of these particles may be quite regular in some mitochondria or seemingly haphazard in others, and there is some question as to their function (see later). A square micron ($1 \mu^2$) of membrane surface would contain approximately 4,000 particles.

Mitochondria do not possess a constant shape, but can swell and contract in a manner that indicates that these movements are correlated with their chemical activity. In fact, there appears to be present

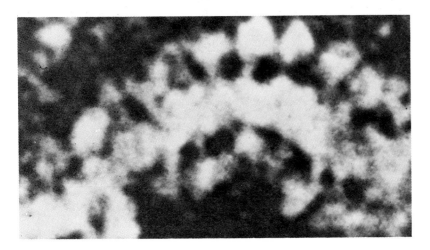

Figure 3.3 *A fragment of the inner membrane of a mitochondrion, showing the stalked particles. As many as 20,000 particles may be present in a single mitochondrion, and they apparently function in the energy-conversion system.*

THE STRUCTURE OF CELLS:
CELL PARTICULATES

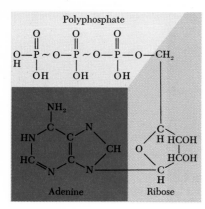

Figure 3.4 Chemical formula of adenosine triphosphate (ATP), with its three components of adenine, ribose, and polyphosphate indicated. The two terminal bonds in the polyphosphate indicated by the wavy line (~) yield large amounts of energy when broken by chemical action.

in the outer membrane a protein similar to actomyosin which is the contractile protein of muscles, and which may be responsible for the volume changes of mitochondria that have been observed. An effective swelling agent is thyroxin, which also increases the general body metabolism of mammalian species as well as the chemical activity of the mitochondria.

It has recently been demonstrated that mitochondria also contain deoxyribose nucleic acid (DNA). As will be discussed later, this is the major hereditary component of cells localized in the chromosomes of the nucleus, but the fact that it is found in mitochondria, and that it differs in molecular nature from nuclear DNA, indicates that, at least to some extent, mitochondria govern their own heritable qualities.

The mitochondria are the aerobic respiratory centers of the cell. This is a topic dealt with in detail by another volume in this series,[*] but because of considerations of structure and function, we will deal with it in brief form here. The mitochondrion is basically an energy converter. The carbohydrates and fats of the cell, together with protein to a lesser extent, are the sources of metabolic fuel used by the cell. Any molecule possesses stored energy in the bonds that link atoms together. When the bonds are broken, energy is released. This happens when a match is lit; all of the bonds are suddenly broken and energy escapes in the forms of light and heat. A cell, however, functions within a limited range of temperatures, and the release of energy from molecules occurs in such a way that bonds are successively broken, with the energy captured by suitable receptors rather than released as heat. As with the match, the waste products are carbon dioxide (CO_2) and water. The principal energy receptor is adenosine triphosphate (ATP) (Figure 3.4), and its importance in cellular economy lies in the fact that this molecule can give up its energy to any part of the cell where work is being performed. ATP, in fact, is cellular currency, and the cell uses it to get things done in much the manner that we use currency to purchase services.

The transformation of carbohydrates and fats by the mitochondria into useful energy is carried out under the direction of enzymes. These are protein molecules, and they speed up such chemical reactions as the breakdown or synthesis of molecules, but are themselves not destroyed in the process. Some 70 or more enzymes are known to be present in the mitochondrion. Some of these are grouped into several enzyme complexes, or assemblies, that govern the processes of *respiration* and are inseparable from the membranes of the mitochondrion;

[*] W. D. McElroy, *Cell Physiology and Biochemistry* (2nd ed.) (Englewood Cliffs, N.J.: Prentice-Hall, Inc., 1964).

others are released in solution when the mitochondrion is ruptured, and are presumed to be in the matrix. Respiration occurs in the presence or absence of oxygen. When oxygen (O_2) is present, a molecule of glucose, for example, will give up its energy to yield 38 molecules of ATP; in the absence of O_2 only 2 molecules of ATP are formed, and the end product is ethyl alcohol (fermentation) rather than CO_2 and water, or lactic acid if the cell is of animal origin.

The transfer of energy from carbohydrates, fats, and proteins into ATP, governed by the numerous enzymes of the mitochondrion, could not be carried out efficiently unless the enzymes were grouped in such a way as to form assembly lines. That is, the breakdown or synthetic product of one enzyme becomes the substrate upon which the next enzyme in line acts. It now appears that most, if not all, of these enzymes are bound to or are an integral part of, the membranes of the cristae. A possible arrangement of some of the enzyme assemblies is indicated in Figure 3.5, and a rat liver mitochondrion is thought to possess as many as 20,000 of these assemblies. The stalked particles shown in Figure 3.3 may also be part of enzyme complexes, but since the size of the particles is too small to accommodate all of the enzymes in a single particle, uncertainty still exists as to the precise significance of the particles to the enzyme complexes.

In any event, it is clear that the mitochondrion is a remarkably complex structure concerned with the capture, conversion and transfer of cellular energy. Its principal function is to form and release ATP for use by the rest of the cell, but it must also, via its membrane structure and enzyme systems, govern selectively the transport of material in and out of itself.

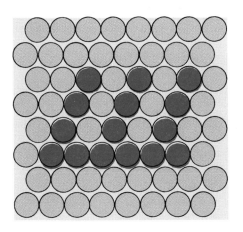

Figure 3.5 *Schematic drawing of a suggested arrangement of the phosphorylating enzymes (shown in solid color) in an "assembly line" and embedded in the protein wall (shown by lighter colored circles).* (Courtesy of Dr. A. H. Lehninger.)

When a cell divides, the mitochondria are distributed to the daughter cells in roughly equal numbers. As the cell enlarges, the mitochondria divide to bring their number up to the level required for proper functioning. It has been suggested that mitochondria may also be formed *de novo* by invaginations of the cell membrane (as indicated in Figure 2.4). The hypothesis, however, fails to account for enzyme systems that are peculiar to mitochondria and for the presence of mitochondrial DNA. At the moment, it seems more likely that mitochondria arise only from preexisting mitochondria, and by division. It is also perhaps likely that mitochondria are symbiotic "cells," possibly of bacterial origin, residing in the cytoplasm of another cell; that is, that they are actually invaders which have become not only established but also essential to the host cell. During the course of evolution, according to this hypothesis, their role in the symbiotic economy has become refined to the point where their function is now crucial.

LYSOSOMES

When mitochondria are sedimented out of a solution of burst cells, they are often contaminated by a structure similar in size to themselves. These have recently been identified as lysosomes. Like mitochondria they possess an outer limiting membrane; they lack, however, the internal cristae and perform a different cellular function. Their internal morphology is very variable, for they can be seen to contain fragments of mitochondria and other membrane systems, pigment

Figure 3.6 **Suggested origin for lysosomes from vesicles derived from the Golgi complex.** *The vesicles (a) surround cellular materials, (b, c) coalesce, and then (d) digest the enclosed materials by means of hydrolytic enzymes.* (Modified from Dr. D. Brandes.)

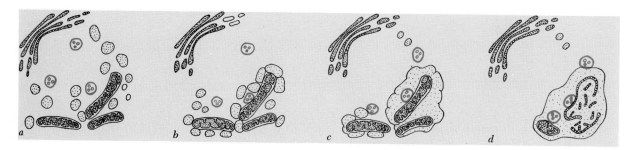

granules, and other electron-dense bodies. There is a strong suggestion that they have their origin from vesicles derived from the Golgi complex (Figure 3.6). This would provide a means of rapid increase in numbers. If this is their origin, it is unlikely that they are related to mitochondria in function, structure, or origin.

The contents of lysosomes provide a clue to their function, for it would appear that these structures are in the process of being broken down, and consistent with this idea is the presence in lysosomes of a number of lytic enzymes, that is, those enzymes which can break down and destroy cellular fragments and large molecules. Lysosomes therefore are the disposal units of the cell, removing foreign bodies or elements of cellular architecture no longer needed.

The lysosomes also seem to increase in number in cells destined to break down and die. As we shall discuss in Chapter 10, cell death and replacement are part of normal developmental processes, and the lysosomes provide the destructive enzymes needed for the dissolution of such cells. Presumably the containment of these enzymes within a lysosome provides a means of selective destruction; rupture of the membrane releases these into the body of the cell, and cytolysis occurs.

PLASTIDS

The plastids of the plant cell share with mitochondria the major problems of energy conversion within the cell. The sun, of course, is the ultimate source of all energy needed in the maintenance and continuation of life. This energy arrives at the surface of the earth as heat and light. Since the cell can function only within a limited range of temperature, heat energy cannot be used to any major extent, but light energy can. The cell, through long ages of change, has evolved the machinery to do this job.

The sequence of the trapping of light energy, its conversion into chemical energy, and its storage in molecules derived from CO_2 and water, is known as *photosynthesis*. This process is described elsewhere in this series;[*] here we need only note that photosynthesis is initiated by the capture of light energy through absorption in the green pigment *chlorophyll*. The *chloroplast* is the cytoplasmic particle in which this takes place.

Electron microscopy reveals that the chloroplast is a structure of

[*] W. D. McElroy, *Cell Physiology and Biochemistry* (2nd ed.) (Englewood Cliffs, N.J.: Prentice-Hall, Inc., 1964); A. W. Galston, *Life of the Green Plant* (2nd ed.) (Englewood Cliffs, N.J.: Prentice-Hall, Inc., 1964).

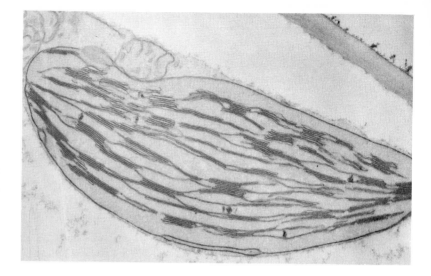

Figure 3.7 Electron micrograph of a chloroplast from the duckweed, Lemna minor. *An outer membrane surrounds the structure, and the grana do not consist of many layers (compare with Figure 3.8). (Courtesy of Dr. H. J. Arnott.)*

Figure 3.8 Portion of a chloroplast from the Spanish bayonet, Yucca. *The grana here consist of a great many layers of membranes (compare with Figure 3.7), with the stroma relatively reduced in amount. This specimen is negatively stained with osmium tetroxide, contrasting therefore with the positive image in Figure 3.7. (Courtesy of Dr. H. J. Arnott.)*

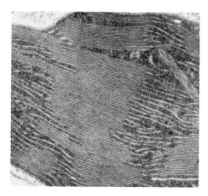

considerable complexity (Figures 3.7 and 3.8). It is bound externally by a unit membrane, and is organized internally into a series of lamellar areas (*grana*) which are connected to each other by extensions of the lamellar structures. The intergrana areas are referred to as the *stroma*. The grana can be visualized as pieces of many-layered plywood, or as stacks of coins, lying in a less well-organized stroma. Within the grana, the chlorophyll molecules are oriented in a monolayer sandwiched in between layers of enzymatic proteins and intimately associated with lipids and carotenoids (Figure 3.9), an arrangement that makes for efficiency not only in the trapping of light energy, but for its conversion, conduction, and utilization in photosynthesis. The stroma is best thought of as the aqueous part of the chloroplast, containing dissolved salts and enzymes, but as Calvin's diagram suggests, enzymes are also part of the lamellar structure of the grana. As Figure 1.4 indicates, some of these enzymes can be extracted intact, and as particulate structures.

Any disturbance induced in the lamellar structure of a granum leads to a reduction in its efficiency. For example, when the unicellular alga *Euglena* is grown in the dark for long periods, it loses its green color; when this happens the interior of the chloroplast breaks down and the lamellae disappear. When this organism is exposed to light again, the formation of lamellae and chlorophyll proceeds simultaneously. Within 4 hr, thin lamellae can be seen forming; by 72 hr, the fully formed chloroplast is evident.

From this kind of information we can conclude that an ordered lamellar structure is necessary for the entire process of photosynthesis to occur; structure and function are interrelated. Biologists have long hoped that they could induce solutions of chlorophyll to photosynthesize in a test tube, and thus provide a method for the efficient production of sugars. It seems likely, though, that they will have to find some way to arrange the molecules in an ordered sequence before they can accomplish this since, at the present time, photosynthesis has been induced only when intact chloroplasts are used.

Chloroplasts may assume many forms, and vary widely in number per cell, in different plants. In some algae, such as the filamentous *Spirogyra*, only a single spiral chloroplast is present in each cell; when the cell divides, it divides at the same time. In contrast, a cell in the spongy part of a grass leaf may have 30 to 50 chloroplasts; their division, which occurs in the immature, or proplastid, state, is not correlated with cell division in any exact way. The stacked grana are missing in some chloroplasts, as in some brown algae, to be replaced by long membranes running the length of the chloroplast, but these presumably function in the same manner as the grana. The blue-

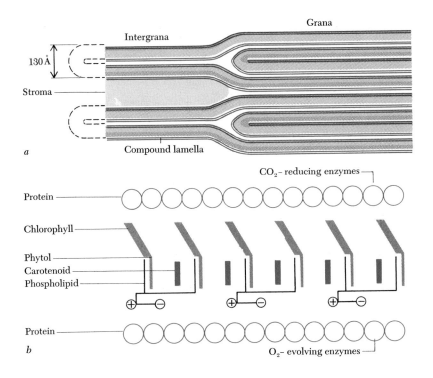

Figure 3.9 *Schematic representation of the layered arrangement of lamellae in the grana (a) and the possible arrangement of molecules in a lamella (b). The enzymes involved in photosynthesis and photophosphorylation are part of the protein layers and may be extracted as particulate structures that can function after the chloroplast has been disrupted. The carotenoid and phospholipid molecules assist in the transfer of energy captured by the chlorophyll. (After A. J. Hodge and M. Calvin.)*

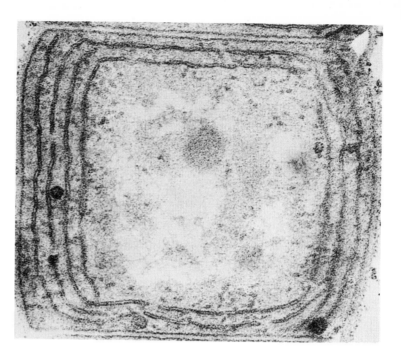

*Figure 3.10 **Electron micrograph of a cell of the blue-green alga,** Oscillatoria. The photosynthetic membranes are loosely layered around the peripheral portion of the cell; the lighter central region containing numerous fine filaments is the nuclear area.* [From W. T. Hall and G. Claus, *Journal of Cell Biology,* 19 (1963), 551–564.]

*Figure 3.11 **Electron micrograph of the marine bacterium,** Nitrosocystis oceanus. The elaborate membrane system is comparable to the photosynthetic membrane of plastids, but the former, however, is engaged in chemosynthesis, a process in which energy is obtained from the alteration of chemical compounds rather than from light sources.* (Courtesy of Dr. S. Watson.)

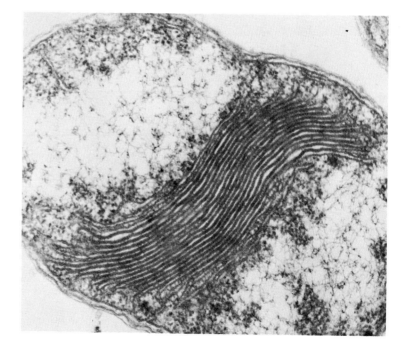

green algae, on the other hand, lack definite chloroplasts; instead they possess loosely arranged membranes in the cytoplasm on which the photosynthetic pigments are layered (Figure 3.10). Only in bacterial cells do we find a photosynthetic capacity unassociated with an obvious membranous structure. However, the vacuolelike *chromatophores*, which are the photosynthetic units, are bounded by membranes, but we know little about the molecular arrangement of the light-absorbing pigments. However, bacteria kept in the dark lose their chromatophores and are no longer photosynthetic; the chromatophore thus behaves as the chloroplast of *Euglena*, and is its functional, but not structural, equivalent. Other bacteria, on the other hand, may have beautifully organized layers of membranes comparable to grana, but in that depicted in Figure 3.11, the membranes are chemosynthetic rather than photosynthetic, the energy for synthesis being derived from the breakdown of sulfur compounds.

All plastids do not contain chlorophyll and function photosynthetically. Some, as in the potato tuber, are for starch storage (Figure 3.12); others may contain oil or protein. These lack the lamellar construction of the chloroplast. However, they are derived from a common type of plastid in the cell and what each becomes depends upon the

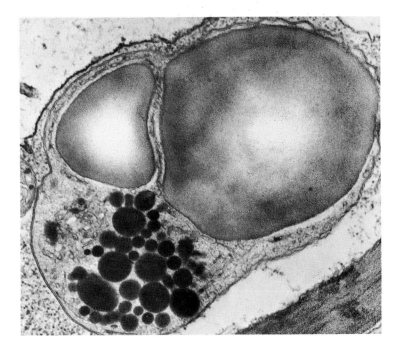

Figure 3.12 **A starch-storage plastid from the tuber of the sweet potato. The large light masses are stored starch; no grana are present.** (Courtesy of Dr. H. J. Arnott.)

kind of cell in which it is found at maturity. In the tomato, for example, the green chloroplast can be seen to transform itself into a red chromoplast as the fruit ripens.

Most students who have a nodding acquaintance with biology think of photosynthesis only as the utilization of light energy to reduce CO_2 to simple sugar molecules, with the necessary hydrogen atoms being derived from water. But there is more to it than that. We now know that the light energy trapped by chlorophyll can also be funneled, through a series of enzymatically controlled reactions, into ATP, which is utilized for completion of the conversion of photosynthetic products into carbohydrates.[*] The chloroplast is, therefore, a dual energy converter, since the energy of sugars and ATP can be utilized by the cell in a variety of ways.

Plastids, like mitochondria, do not arise *de novo*, but from other plastids by division, and most likely at the *proplastid*, or undeveloped, stage. In some species, however, as in *Spirogyra*, plastid division is correlated in time with cell division, and it is the mature, fully formed plastid which divides into two parts. In higher forms, as in the ferns and seed plants containing numerous plastids per cell, this correlation is not evident, and it is believed that an increase in plastid number is by division of the proplastid.

As with mitochondria, plastids also possess their own DNA, and the suggestion has been made that they have been derived from symbiotic invaders which, during the course of evolution, have taken over an essential part of cellular metabolism. They thus form a somewhat independent unit of inheritance within the cell even though it is evident that their function and morphology is determined, in part, by the nucleus. The development of plastids is, therefore, of interest for it reflects the interaction of light and heritable factors in determining structure. A plant, grown from seed in the dark, lacks green color; it will turn green rapidly on exposure to light. In the dark, only proplastids are evident. They are small in size and contain a *prolamellar body* which has the appearance of a lattice structure. In the presence of light, the prolamellar body becomes rearranged into a series of tubelike structures which grow in length and finally become organized into the grana and intergrana layers (lamellae) characteristic of the species. As the lamellae and grana are forming, the precursor molecules of chlorophyll become altered into functional chlorophyll and layered on the forming membranes. Only then can the photosynthetic apparatus function properly. Since the whole pro-

[*] Discussed more fully in W. D. McElroy, *Cell Physiology and Biochemistry* (2nd ed.) (Englewood Cliffs, N.J.: Prentice-Hall, Inc., 1964).

cess consists of a number of steps, interruption by environmental or genetic factors can lead to various plastid abnormalities, many of which have been studied because of their changed morphology and function.

Electron microscopy has revealed that ribosomes are found in every kind of cell, and this is to be expected since it has been shown biochemically that they are intimately concerned with protein synthesis (see Chapter 4). They are generally spherical, and have a diameter of about 200 to 250 Å. Composed of approximately 60 percent RNA and 40 percent protein, they have been shown to consist of two unequal subunits, each of which contains an RNA molecule of particular size, and probably particular proteins as well since they are orderly in shape and size. Aggregation of ribosomes frequently occurs, and in cells which synthesize a particular kind of protein, the aggregates, or *polysomes,* consist of definite numbers of ribosomes (Figure 3.13*b*). The reason for this is that when a protein of particular size is being synthesized, the surface area of a single ribosome is apparently insufficient for the reactions to take place, and several ribosomes are bound together by strands of additional RNA.

Cells actively engaged in protein synthesis may have the ribosomes attached to the outside membranes of the ER, or they may float free in the cytoplasm. Although too small to be seen in the light microscope, the presence of ribosomes may be detected by the fact that they stain readily with a basic dye such as erythrocin.

Formation of ribosomes is initiated in the nucleus, specifically in the nucleolus, their site of origin, but they are assembled in the cytoplasm (Figure 3.14). Their presence in mitochondria and plastids has already been discussed. Their likely role in the latter two particulate structures is also one involved in protein synthesis; however, this synthesis is of a more localized nature, the proteins being utilized within the plastid or mitochondrion rather than in the remainder of the cytoplasm (where highly specific proteins such as enzymes are formed).

Characteristic of animal cells and of ciliated plant cells is a cylindrical structure about 3,000 to 5,000 Å in length, and about 1,500 Å in diameter. This is the centriole. It is found at the edge of the nucleus—often in an indentation of the nuclear membrane—and is also often

Figure 3.13 **Diagram of two ribosomes (a) and aggregates of ribosomes, or polysomes (b). The aggregates from particular cells fall into definite size classes, indicating that the number of ribosomes per cluster is determined by the length of the mRNA holding them together.**

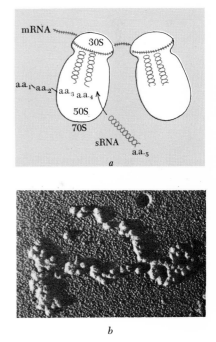

b

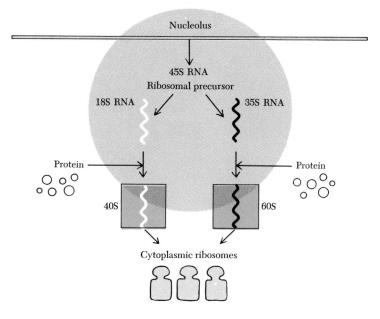

Figure 3.14   Schematic representation of the origin and formation of cytoplasmic ribosomes. The S values stand for particles which sediment out in an ultracentrifuge at particular speeds and which can be separately collected and analyzed. The chromatin at the nucleolar organizer region forms 45S ribosomal precursor, which separates into two units of unequal size, 18S and 35S. These combine with protein, presumably from outside the nucleolus, to form 40S and 60S ribonucleoprotein particles, which, after leaving the nucleolus, unite to form the ribosomes found in the cytoplasm. (After R. P. Perry.)

associated with, and partially surrounded by, the Golgi complex. When two centrioles are found together, one is generally oriented perpendicular to the other. Their role in cell division is critical in determining the axes of division (page 98).

When cut in cross section, the centriole shows nine sets of triplet tubules arranged in a circle (Figure 3.15), and running the length of the hollow cylinder. The inner of the triplet tubules is connected to the outer triplet of the adjacent set, and arranged in such a manner that the cylinder forms a long helix of interconnecting strands.

The centrioles have been said to contain DNA, and consequently to be self-replicating, but their chemical nature and exact function remain to be determined. It is believed that they replicate by a process of budding. In ciliated cells they replicate repeatedly to give a number of copies which migrate out to the cell surface where they become the *basal bodies* which produce, and presumably govern the action of, cilia or flagella. The central cylinder of a cilium reflects the tubular arrangement of the centriole, but with the variations noted in Figure 3.15. They also divide well in advance of the cell, as can be seen very clearly in the huge centrioles found in some species of protozoa (Figure 3.16).

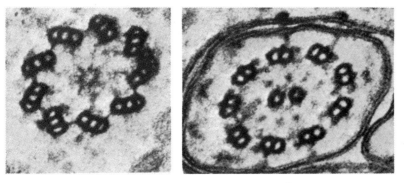

The cells of flowering plants lack a centriole, and in the ferns and gymnosperms (pines and their allies), this structure is found only in the sperm cells, where it forms the flagella. This would suggest that the centriole arises *de novo* in these cells, but until the exact mode of replication is determined this remains only a possibility.

Vacuoles are more characteristic of plant than of animal cells. They are liquid-filled areas of the cell, and are surrounded by a unit membrane, the *tonoplast*, which prevents the dilution of the cytoplasm by

VACUOLES

Figure 3.16 **Formation of the spindle by the long centriole in the proto-zoan** Barbulanympha. **Note that the new centrioles for the next division are already in evidence; these will lengthen between divisions to reach full size. The rounded structure is the centrosome.** (From L. R. Cleveland.)

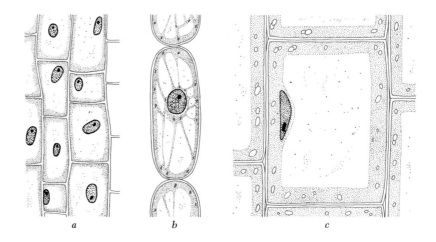

Figure 3.17 **Three examples of plant cells, illustrating the manner by which the formation of a vacuole pushes the cytoplasm to the outside, thus increasing the exchange of materials between the cytoplasm and the exterior of the cell. (a) Cells from the one-cell-thick skin in an onion bulb. (b) Cell from the stamen hair of the spiderwort,** Tradescantia. **(c) Typical cell of a leaf, showing the plastids, nucleus, and cytoplasm forced to the outside. Strands of cytoplasm may often cross through the vacuole, as shown in (b).**

the contents of the vacuole. The tonoplast, while of unit construction similar to that of the cell membrane, has its own characteristic degree of permeability, and can thus preserve areas of segregation within the cell. For example, salts, acids, sugars, and pigments are frequently found here, and often in such concentrations as to form large crystals. The red pigment of beet roots as well as that of many flowers is found in vacuoles, while the degree of acidity of the vacuole may be very different from that of the cytoplasm.

The dissolved materials in the vacuole maintain the proper internal pressure in the cell; water moves into the hypertonic vacuole, and the tonoplast keeps the dissolved materials from moving out. The cell thus maintains its turgidity. The vacuole also provides a dumping place in the cell for waste or unwanted materials.

In young, actively growing cells the vacuoles are small, but eventually they coalesce to form one large vacuole which forces the cytoplasm to the outside of the cell (Figure 3.17). There are obvious advantages to such a system. A ready exchange of gases can take place in a thin layer of cytoplasm, which is clearly necessary in a cell engaged in photosynthesis; the molecules of O_2 and CO_2 do not have to pass through a mass of cytoplasm, and the thin layer forces the chloroplasts out to the surface of the cell, where light can readily reach them.

Electron microscopists have recently recognized that tubules and filaments in the cytoplasm form a significant structural element of cells. The two types are recognizably different from each other. The tubules appear hollow and are about 250 Å in diameter with walls about 60 Å in thickness. They appear to be similar to the individual tubules that are aggregated to form centrioles and cilia. The filaments, on the other hand, are solid rods 40 to 50 Å in diameter. Both are of indefinite length. They occur in particular types of cells and probably serve different functions in addition to their role as structural elements.

Microtubules are commonly aggregated to form the spindle of dividing cells (Figure 3.16), where they extend from chromosome to pole, or from pole to pole. They function somehow in the movement of chromosomes during cell division. In other cells such as neurons, microtubules are arrayed in the long axis of cell where they may serve as a cytoskeleton to give a degree of rigidity to the tenuous neuron. In plant cells, the microtubules are extended along strands of cytoplasm, and it is along this same orientation that the cellulose fibers are laid down during cell-wall formation. The tubules may simply be oriented by the streaming cytoplasm, but it is more likely that they play a positive role in laying down the cell wall.

Filaments are probably of many kinds, but all seem to be proteinaceous in character. Those found in contractile muscle will be discussed later, but those commonly present in mammalian epidermal cells are often aggregated into substantial masses of *tonofibrils*. When cut in cross section, they can be mistaken for small ribosomes, but their diameter (40 to 50 Å) is only one-third to one-fourth that of normal ribosomes.

A wide variety of cell products are found in the form of particulates. Some of these are waste products that the cell immobilizes as insoluble crystal structures; the crystals often found in the vacuoles of plant cells would be examples of these. Others are pigment masses such as the melanin found in certain cells (melanocytes) of the skin or liver, while still others are substances stored for eventual use by the cell—

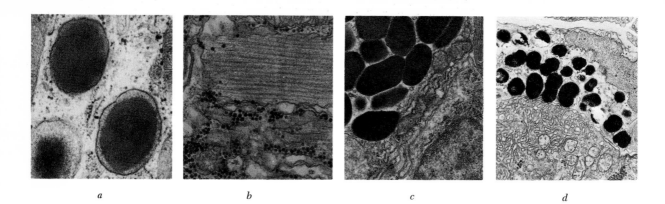

Figure 3.18   Electron micrographs of inclusions found in animal cells.
(a) Zymogen granules in a pancreatic cell. (b) Small glycogen granules found
between the sarcomeres of a muscle cell. (c) Mucus droplets in the branchial
epithelial cell of a fish. (d) Pigment granules in a melanocyte.

glycogen in the liver or muscle, lipids in the fatty tissues—or destined
for transport out of the cell—mucus in the goblet cells of the stomach
or intestine, zymogen granules in the pancreatic cells. Several of these
are illustrated in Figure 3.18; their identification is made on the basis
of morphology and staining reactions.

THE MOST PROMINENT PARTICULATE FEATURE OF A CELL WHEN VIEWED under the microscope is the nucleus (Figure 4.1). It is an almost universal structure of cells at some time during their life cycle, although such cells as the sieve tubes of vascular plants and the red blood cells of mammalian vertebrates may lose their nuclei by the time they are fully differentiated. The only exceptions to the statements above are found among the cells of bacteria and blue-green algae, which possess a loosely defined *nuclear area* rather than the definitive, membrane-bound nucleus of higher plants and animals.

A single nucleus per cell is also a general rule for the vast majority of known cells. A uninucleate cell appears, therefore, to be the most efficient and economical manner of packaging living substance. Once again, however, exceptions are known. Striated muscle cells of mammals and latex tubes of vascular plants, for example, are generally multinucleate, although these may have achieved multinuclearity by cell fusion rather than by divisions of the nucleus into two or more parts without division of the cytoplasm. Many fungi are also multinucleate, for their strands of mycelia often lack the walls which

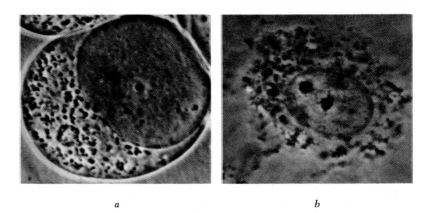

Figure 4.1 **Nuclei.** *(a) The very large nucleus of an amphibian spermatocyte about to enter meiotic division.* (Courtesy of Dr. J. Kezer.) *(b) An animal fibrocyte growing in tissue culture.*

a   b

break a mass of cytoplasm into cellular structures. Also, *Paramecium*, the well-known protozoan, is regularly binucleate, having a *micronucleus* that functions in an hereditary capacity and a *macronucleus* that governs the metabolism of the organism.

In appearance, the nucleus is generally a rounded body, although in vaculated plant cells it may be lens-shaped, as the result of pressure from the vacuole, or lobed as in some blood cells (Figure 4.2) and in cells of the spinning gland of a spider. The reason for the lobate form is not known, but presumably this provides the nucleus with a greater surface area per unit of volume.

Figure 4.2 **Electron micrograph of the highly lobed nucleus found in a human polymorphonucleated leucocyte (white blood cell).**

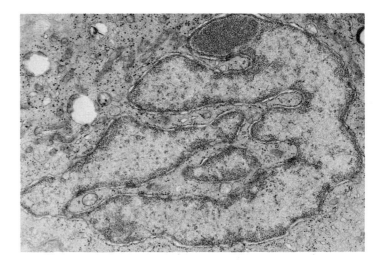

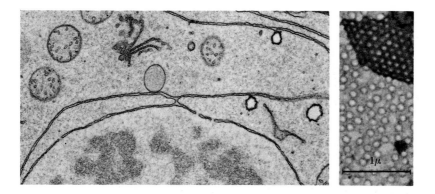

Figure 4.3 *Electron micrographs of nuclear membranes.* (a) *Portion of a plant cell with the nucleus and shadowy chromosomes at the bottom; the double nature of the nuclear membrane, the pores, and the connection of the outer portion of the nuclear membrane with the ER can be seen.* (Courtesy of Dr. G. Whaley.) (b) *A face-on view of the nuclear membrane from a frog's egg, showing the regular arrangement of the nuclear pores; the darker portion is an overlying membrane fragment.* (Courtesy of Dr. R. W. Merriam.)

Figure 4.4 **Diagram showing the character and dimensions of the nuclear pores, as seen in amphibian oocyte nuclei.** (After Dr. J. D. Gall.)

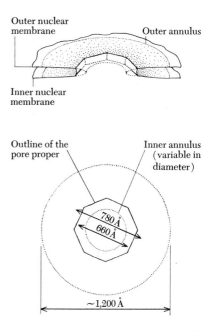

A double-layered *nuclear membrane* surrounds the nucleus, with the two membranes being separated by a *perinuclear* space of varying width. As Figure 4.3 shows, the outer membrane is often continuous with elements of the endoplasmic reticulum, and like the ER it often bears ribosomes on its outer, or cytoplasmic, surface. It seems quite likely, in fact, that the outer nuclear membrane, and possibly the inner one as well, is derived from the ER, since during cell division the newly formed nuclei have their membranes partially formed from coalescing fragments of the ER.

Nuclear membranes also display an additional structural feature in that the membrane is periodically interrupted by pores, octagonal in shape, and at the edges of which the inner and outer membranes are continuous. The pores may be very regular in distribution, as in amphibian oocyte membranes, or they may be irregularly spaced, as in plant cells (Figure 4.3). In animal cells, at least, the pores are also highly regular in shape and size (Figure 4.4). The fact that pores are present in most, if not all, nuclear membranes would suggest that they represent avenues of exchange between nucleus and cytoplasm. However, high-resolution electron microscopy of nuclear pores, particularly in animal cells, shows that the pore is not necessarily a sim-

49

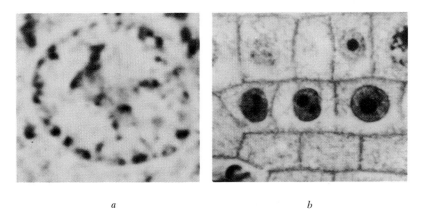

Figure 4.5   Nuclear structures. (a) An animal cell in interphase, showing the heterochromatic masses that tend to lie against the nuclear membrane as well as being located centrally; the two dark spots outside the nuclear membrane at the top center are the paired centrioles. (b) A group of root-tip cells of the onion in interphase, showing heterochromatic masses as well as nucleoli. The euchromatin is lightly stained and not readily visible.

a   b

Figure 4.6   A meiotic cell (microsporocyte) in rye, showing the attachment of the nucleolus to a particular pair of chromosomes (the pair centrally located). (Courtesy of Dr. R. A. Nilan.)

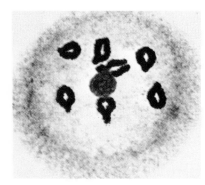

ple opening between the two membranes, but is often filled with a "plug" that would tend to block passage of substances in and out of the nucleus. The significance of the pores, therefore, is unsettled, as is their mode of origin, although it has been suggested that their octagonal shape stems from the regular circular aggregation of eight microtubules.

The nucleus stains readily with basic dyes after acid fixation to reveal a network of fine threads, among which coarser masses of stained material stand out (Figure 4.5). The electron microscope, despite its great powers of resolution, reveals the same structure with no additional details. This is the *chromatin,* with the fine threads being *euchromatin,* the coarser material *heterochromatin.* In addition, the nucleus contains a *nuclear sap,* and one or more dense bodies, or *nucleoli* (singular, *nucleolus*). The latter are formed by particular chromosomes that possess a structure known as the *nucleolar organizer* (see Figure 4.6), which manufactures nucleolar material and organizes it into a dense body. The character of the nucleolus varies with the kind of cell and its metabolic state; it is larger and more dense in active cells, rapidly growing embryonic cells, or those engaged in protein synthesis than in those that are relatively inactive metabolically. Electron microscopy also reveals an internal differentiation in the nucleolus in the form of a loose network of strands—the *nucleolonema* (Figure 4.7)—made up of granular material, probably ribosomes, of about 150 Å in diameter, and a less structured area, the *pars amorpha.* The significance or role of these two nucleolar features is not yet known, but as will be discussed later and as depicted in Figure 3.14, the nucleolar organizer region of the chromosome is not only responsible for the formation of nucleoli, but is also the site of

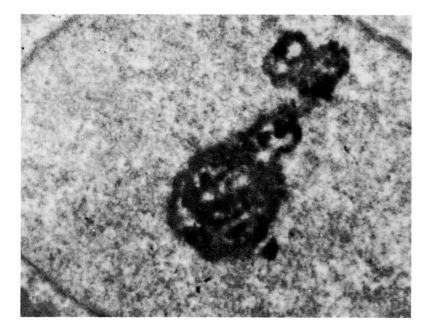

Figure 4.7 **The nucleus of a mouse fibrocyte,** *showing the fibrous nature of the nucleolus. The dark strands are the nucleolonema; although not evident in this illustration, there is often a distinct clear area in the nucleolus called the* pars amorpha. (Courtesy of Dr. E. Borsyko.)

formation of ribosomes which, accumulating in the nucleoli, eventually pass to the cytoplasm to become organized into the protein-synthesizing machinery of the cell. This is indicated by the fact that radioactive *uridine,* a nucleotide which becomes incorporated into the RNA of ribosomes, is first concentrated in the nucleolus, after which it then passes to the cytoplasm.

NUCLEUS AS A CONTROL CENTER

The mammalian red blood cell lacks a nucleus when fully differentiated. It is also a cell restricted in metabolic activity, limited in growth and length of life (120 days), and incapable of further division. From these observations, the almost universal occurrence of a single nucleus per cell, and the fact that in cell division the nucleus alone goes through a dramatic and exact partitioning of its chromatin content, we can draw the conclusion that the nucleus is an essential structure of cells. The fact that it is also the control center of the cell may be judged from the results of two kinds of experiments. In an ameba, for example, it is possible to remove the nucleus by means of surgery or with a microneedle or micropipette. The cytoplasm is not damaged by such surgery, and the cell can metabolize for several days or even weeks

THE STRUCTURE OF CELLS:
THE NUCLEUS

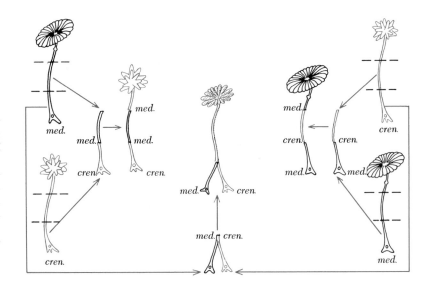

Figure 4.8 **Influence of the nucleus on development in** Acetabularia. *Stalk segments of* A. mediterranea *grafted onto nucleus-containing rhizoids of* A. crenulata, *and vice versa produce caps characteristic of the species contributing the nucleus. When two nucleus-containing rhizoids are grafted together, the cap consists of loose rays, as in* A. crenulata, *but their points are more rounded, as in* A. mediterranea.

since the requisite cytoplasmic enzymes and organelles are still functioning. The cell eventually runs down, however, and is unable to rejuvenate itself unless another nucleus is put back into the cytoplasm. Therefore we can view an enucleated cell as a cell without a future, and the nucleus as a necessary organelle providing information or parts to the cytoplasm to keep it functioning properly for an indefinite period.

The nucleus is also the source of information that governs the morphology of cells. We might have assumed this from the distinctive shapes of many unicellular organisms, and from the many kinds of cells found in multicellular organisms, but experiments done by a German biologist, Max Hämmerling, on *Acetabularia,* a single-celled green alga of warm marine waters, demonstrate this beautifully. Two species of *Acetabularia* differ in the shape of their caps (Figure 4.8). If a cap is cut off, it will form again; it is also possible to graft a piece of the stalk of one species onto the decapitated, nucleus-containing *rhizoid* of the other species. When the cap forms again, it is characteristic of the species that contributed the nucleus and not that of the species that contributed the stalk and its contained cytoplasm. When rhizoids of two different species, both containing nuclei, are grafted together, the cap that forms is intermediate in morphology, reflecting the influence of both nuclei. In some manner, then the nucleus causes a cell to do as it dictates, and this control is exercised not only morphologically but functionally as well.

Cellular control must be exercised through chemical means; there is no other way of control. Consequently, a great deal of effort has been devoted to the chemical analysis of chromatin since it is the most conspicuous, and in fact the only universal, element in all nuclei. Such analyses reveal the presence of four major components: DNA, RNA, a low-molecular-weight protein called *histone,* and more complex *residual protein.* We have as yet no clear idea of how these molecules are grouped together to form chromatin, but the RNA and residual protein vary in amount from one kind of cell to another in the same organism, depending on the metabolic activity of the cell. The amounts of DNA and histone per nucleus, on the other hand, are remarkably constant within the cells of the same organism although they differ between species (Table 4.1). The constancy of the DNA-histone complex is what we might expect in the case of a stable controlling system.

A number of experiments have demonstrated that the controlling molecule of the nucleus is DNA. Consider, for example, two strains of bacteria, one resistant to an antibiotic, the other susceptible to its action. Resistance or susceptibility to antibiotics is a heritable trait. If the DNA of the resistant strain is extracted in pure form, and with no other contaminating molecules, this can be added to the culture medium in which the susceptible strain is growing. Some of

Table 4.1   **DNA–haploid complement weight** [a]

Mammals		Reptiles		Fish	
Man	3.25	Snapping		Carp	1.64
Beef	2.82	turtle	2.50	Shad	0.91
Rat	3.4	Alligator	2.50	Lungfish	50.0
Dog	2.75	Water snake	2.51	Miscellaneous	
Mouse	3.00	Black snake	1.48	Maize	8.4
Marsupial	4.5	Amphibians		Drosophila	0.085
Birds		Amphiuma	84.0	Aspergillus	0.043
Chicken	1.26	Necturus	24.2	Neurospora	0.020
Duck	1.30	Frog	7.5	E. coli	0.0040
Goose	1.46	Toad	3.66		

[a] From C. P. Swanson, T. Merz, and W. J. Young, *Cytogenetics* (Englewood Cliffs, N.J.: Prentice-Hall, Inc., 1967), p. 177.

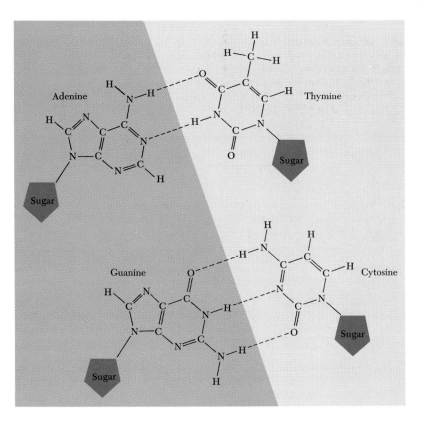

Figure 4.9 *Chemical configurations of the four bases found in the DNA molecule, arranged as base pairs. Thymine and cytosine are pyrimidines; adenine and guanine are purines.*

Figure 4.10 *Schematic and flattened arrangement of phosphates, sugars, and bases to form the DNA molecule, with the right- and left-hand portions of the molecule held together by hydrogen bonds. This arrangement, twisted into a helix by molecular forces, gives the configuration seen in Figure 4.11.*

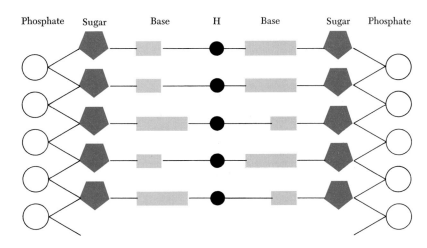

Figure 4.11   The DNA helix, with three different ways of representing the molecular arrangement. (a) General picture of the double helix, with the phosphate-sugar combinations making up the outside spirals and the base pairs the cross bars. (b) A somewhat more detailed representation: phosphate (P), sugar (S), adenine (A), thymine (T), guanine (G), cytosine (C), and hydrogen (H). (c) Detailed structure, showing how the space is filled with atoms: carbon (C), oxygen (O), hydrogen (H), phosphorus (P), and the base pairs.

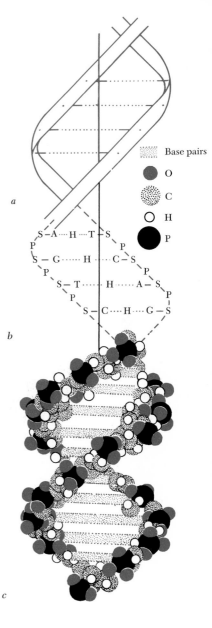

Base pairs

O

C

H

P

a

S—A····H····T—S
  P   P
S—G····H····C—S
   P   P
   S—T····H····A—S
    P   P
     S—C····H····G—S

b

the susceptible cells will be transformed into resistant ones. It has been demonstrated that this transformation is caused by the susceptible strain incorporating into its chromatin a piece of DNA from the resistant form which governs the heritable trait of resistance. DNA is, then, the crucial molecule of inheritance, and its molecular nature becomes a matter of importance, since its structure must be consistent with its function.

Chemical analysis reveals DNA to be a compound of very high molecular weight, and composed of a number of smaller molecules linked together in a precise manner. These include a sugar (*deoxyribose*), *phosphoric acid,* and four heterocyclic bases, two of which are *pyrimidines* (*thymine* and *cytosine*) and two *purines* (*adenine* and *guanine*). The structure of these compounds is given in Figure 4.9 and their arrangement to form DNA is indicated in Figure 4.10. The alternating sugar and phosphate residues form the outside boundaries of the molecule while the base pairs are linked in horizontal planes. The base pairs, however, are not randomly arranged. Adenine and thymine are always linked as a pair, as are guanine and cytosine. Only four possible pairs can therefore be formed: A-T, T-A, G-C, and C-G, and this is determined by the character of the hydrogen bonding between the base pairs. Furthermore, by one of the most dramatic examples of model building to test an idea, Watson and Crick showed DNA to be a double helix, as indicated in Figure 4.11. This structure, therefore, consists of molecular twins, and provides a clue as to how the molecule reproduces itself in preparation for cell division.[*]

We know, for instance, that when a cell divides, the two daughter cells are genetically identical. The mother cell must, therefore, have exactly replicated the hereditary and controlling substance in the cell so that each daughter cell would be similar in genetic content. Figure 4.12 indicates how this is accomplished. The double helix separates

c

---

[*] The fascinating story of the discovery of the structure of DNA is found in *The Double Helix* by J. D. Watson (New York: Atheneum Press, 1968).

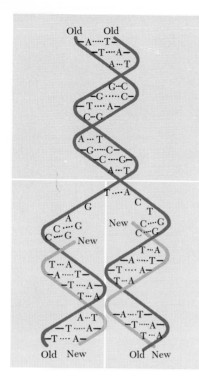

Old   Old

A····T
T····A
A···T

G····C
G   C
T····A
C-G

A···T
G····C
C····G
A···T

T····A   C
G   T
A   New C-G
C····G   C-G
C-G New

T···A   T-A
T···A   A···T
A···T   T····A
T····A   T···A
T···A

A-T   A····T
T····A   T····A
T····A   T···A

Old New   Old New

Figure 4.12 **Replication of the DNA molecule, which takes place during interphase. The old helix unwinds (center) and the two new helices are formed.**

MODE OF CELLULAR CONTROL BY DNA

THE CELL

into its two constituent polynucleotide strands, the molecular twins, by breaking the hydrogen bonds that hold the bases together in pairs. Each strand, as it unwinds, forms a new strand complementary to it, so that at the end of the process of replication two identical double helices are formed. By this mechanism, the hereditary content of all the cells of an organism remain the same, and an oak tree forms only oak cells, and human cells only other human cells.

Organisms differ in the amount of DNA they contain. The DNA in the common colon bacterium, *Escherichia coli*, is a single piece 1,100 μ long, and contains 4.3 million nucleotide pairs. Each nucleotide pair is 3.4 Å long, and there are 10 nucleotide pairs in every complete turn in the helix. The *E. coli* DNA has, therefore, 430,000 turns in its double helix, each of which must be undone when the piece of DNA replicates. By way of contrast, the DNA of a human cell is over 1,000 times greater in amount. It is about 2 m (meters) in length, and contains about 8 billion nucleotide pairs distributed among the 46 chromosomes found in a normal human cell.

If DNA is the crucial molecule of inheritance, and if its structure is described in Figure 4.11, then the arrangement of the nucleotide pairs must provide the information, in some kind of coded form, needed by the cell to carry out its functions in a regular and predictable manner. This emphasizes the logicality of the uninucleate cell and the obvious limitations of a multinucleate condition. A great deal of effort, therefore, has been devoted to breaking the "genetic code," and it is now believed that a solution has been reached.

In 1941, George Beadle and Edward Tatum demonstrated that genes, which govern heritable traits, exercise their control through the medium of enzymes. Thus, if a product X is formed by a series of chemical reactions,

$$A \rightarrow B \rightarrow C \rightarrow D \rightarrow X$$

each step in the series is controlled by an enzyme. If the enzyme controlling $A \rightarrow B$ is missing, or rendered nonfunctional, the remaining steps in the series cannot occur because the product B would similarly be missing.

Virtually every chemical reaction within a cell is governed by enzymes. And enzymes are proteins. Therefore, the uniqueness of every cell and every organism or species is based on the uniqueness of its proteins. If DNA is the crucial molecule of inheritance, then it must in some way determine the specificity of the proteins of the cell. And this specificity must somehow be determined by the sequence of nucleotides in the DNA molecule.

Proteins are similar to DNA because they, too, are macromolecules made up of smaller units strung together in linear sequence. In place of nucleotide pairs, however, proteins are made up of a linked series of *amino acids* (Figure 4.13), and the specificity of action of an enzyme is determined by its sequence of amino acids. But there are 20 amino acids, and only four kinds of nucleotides. Quite obviously, then, if DNA determines the sequence of amino acids in a protein, one nucleotide in DNA cannot determine the kind and position of one amino acid in a protein; such a *singlet* code would take care of only four amino acids. Similarly a *doublet* code is inadequate, since only 16 (4 × 4) amino acids could be handled by the cell in coded fashion. A *triplet* code, however, is more than adequate since 64 (4 × 4 × 4) triplets are possible, and as Table 4.2 indicates, a number of amino acids can be coded by more than one code.

A piece of DNA is, therefore, a coded message which is "read" by the cell, and this message is eventually translated into a particular kind of protein, or enzyme. A message, of course, has a beginning and an end, and although there is doubt about how the message is initiated, it is known that the code provides positive termination points (commas in Table 4.2). However, DNA is in the nucleus, and proteins are formed in the cytoplasm, with the ribosomes playing a role in protein synthesis. Some device must, therefore, carry the message from the nucleus to the cytoplasm where it is then translated into protein formation. A series of brilliant experiments have shown that several kinds of RNA participate in this process. The sequence of events is depicted in Figure 4.14.

DNA is the source of the three kinds of RNA. DNA, therefore, not only replicates itself, but also, by a process called *transcription*, makes the several RNA's. Only one of the two polynucleotides of

*Figure 4.13   **The amino acid sequence of human insulin. The particular** amino acids are abbreviated in this illustration.*

Gly-Ileu-Val-Glu-GluN-Cys-Cys-Thr-Ser-Ileu-Cys-Ser-Leu-Tyr-GluN-Leu-Glu-AspN-Tyr-Cys-AspN

Phe-Val-AspN-GluN-His-Leu-Cys-Gly-Ser-His-Leu-Val-Glu-Ala-Leu-Tyr-Leu-Val-Cys-Gly-Glu-Arg-Gly-Phe-Phe-Tyr-Thr-Pro-Lys-Thr

DNA transcribes, and the RNA that is made is not only complementary to the piece of DNA, but the *thymine* of DNA is replaced in RNA by *uracil*. If a piece of DNA has the following sequence of nucleotide pairs

G	A	T	C	C	A	G	T	C	A	A	T	C	C	A
C	T	A	G	G	T	C	A	G	T	T	A	G	G	T

Table 4.2   *Triplet codes and the amino acids each code determines*

CODE TRIPLETS	AMINO ACID	CODE TRIPLETS	AMINO ACID
AAA	Lysine	CAA	Glutamine
AAG	Lysine	CAG	Glutamine
AAC	Asparagine	CAC	Histidine
AAU	Asparagine	CAU	Histidine
AGA	Arginine	CGA	Arginine
AGG	Arginine	CGG	Arginine? a
AGC	Serine	CGC	Arginine
AGU	Serine	CGU	Arginine?
ACA	Threonine	CCA	Proline
ACG	Threonine	CCG	Proline?
ACC	Threonine	CCC	Proline
ACU	Threonine	CCU	Proline
AUA	Isoleucine?	CUA	Leucine
AUG	Methionine	CUG	Leucine
AUC	Isoleucine	CUC	Leucine
AUU	Isoleucine	CUU	Leucine
GAA	Glutamic acid	UAA	Gap, b
GAG	Glutamic acid	UAG	Gap,
GAC	Aspartic acid	UAC	Tyrosine
GAU	Aspartic acid	UAU	Tyrosine
GGA	Glycine?	UGA	Tryptophan
GGG	Glycine?	UGG	Tryptophan
GGC	Glycine?	UGC	Cysteine
GGU	Glycine	UGU	Cysteine
GCA	Alanine?	UCA	Serine
GCG	Alanine?	UCG	Serine
GCC	Alanine?	UCC	Serine
GCU	Alanine	UCU	Serine
GUA	Valine?	UUA	Leucine
GUG	Valine	UUG	Leucine
GUC	Valine?	UUC	Phenylalanine
GUU	Valine	UUU	Phenylalanine

[a] Question marks indicate uncertainty of code specification at the present time.

[b] Commas represent the end of any message.

and if only the bottom nucleotide strand transcribes, then the RNA made will be

G A U C C A G U C A A U C C A
| | | | | | | | | | | | | | |

The correctness of this interpretation is demonstrated in the following way: if the DNA is separated into its two polynucleotide strands, and then mixed with RNA, specific pieces of RNA will unite with specific pieces of DNA to form "hybrid" DNA-RNA molecules:

```
——————————————— DNA
 C  T  A  G  G  T  C
 G  A  U  C  C  A  G
| | | | | | | | | | | | | | ——— RNA
```

The RNA in ribosomes, called rRNA, consists of two pieces. Both are made by the nucleolar organizer region of the chromosome, as-

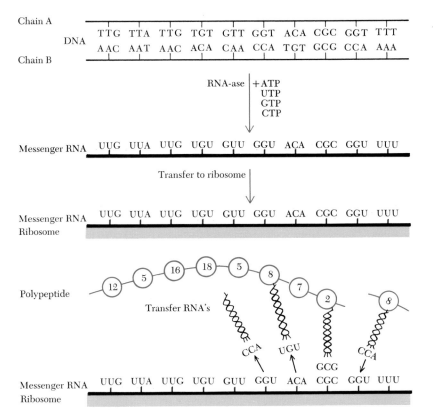

Figure 4.14 *The sequence of events whereby the information coded in the DNA molecule is transcribed into messenger RNA (mRNA) and then, by translation, into the formation of a particular protein. It is assumed here that it is the Chain B of the DNA that is being "read." The mRNA contains the code to be read, while the transfer RNA's (tRNA's) possess an anticode, which matches the code by base-pairing. The tRNA at the right carries an activated amino acid that has not yet been inserted into the growing polypeptide.*

sembled and combined with protein in the nucleolus, and transferred and then organized into ribosomes in the cytoplasm. *Transfer RNA,* or tRNA, is made by other regions of the chromosome. These contain about 80 nucleotides, are all terminated at one end by a given nucleotide sequence, all contain a specific triplet code—called an *anticode*—in the middle of the molecule, and all can be activated so that an amino acid can be attached as indicated in Figure 4.14. Since there are 64 triplet codes, there is theoretically, an equal number of tRNA's possessing the complementary anticodes, some of which have been identified.

The rRNA's and tRNA's assist in protein formation, but the sequence of amino acids in a protein is determined by the sequence of nucleotides in a third kind of RNA called *messenger* or mRNA. Again this is made by the DNA and moves to the cytoplasm where it becomes attached to the ribosomes. As Figure 4.14 indicates, the proper tRNA's transport activated amino acids to the mRNA, with the triplet code of mRNA and the complementary anticode of tRNA determining the order in which the particular amino acids are to be linked into a protein. When the protein is completed, it is released to take its proper place in the cytoplasm where it acts either as an enzyme or as a structural molecule in membranes, plastids, or other cellular parts.

The entire process of RNA and protein formation is governed by enzymes and requires cellular energy for its completion. The process is described in detail in another volume in this series,* but by the brief description given here we can visualize how a gene in a chromosome, consisting of a number of nucleotide pairs, can determine a given trait (the presence of a particular enzyme is a biochemical trait). Therefore, if a protein contains 100 amino acids arranged in a specific order, we know that this must have been initiated by a piece of DNA consisting of 300 nucleotide pairs. It is not yet possible, however, to decode the DNA even if the structure of a protein is fully known. As Table 4.2 indicates, some amino acids have several triplet codes, and we do not yet know the conditions which determine when one or the other is to be used.

* W. D. McElroy, *Cell Physiology and Biochemistry* (2nd ed.) (Englewood Cliffs, N.J.: Prentice-Hall, Inc., 1964).

THE PLASMA MEMBRANE IS GENERALLY CONSIDERED TO BE THE OUTER living limit of the cell, but not necessarily its outer boundary. We can see such outer boundaries most readily in plant cells, many of which possess heavy walls of cellulose, but animal cells and many unicellular organisms also exhibit a number of somewhat comparable external substances, some of which are visible only at the level of electron microscopy, while others are readily discernible with the light microscope. To state categorically, however, that any part of the cell, either inside or on the immediate outside, is living or non-living is to presume to define life and, as everyone knows, doing so has its pitfalls. Nevertheless, what lies at the outer boundary of the plasma membrane determines the discrete character of cells.

Most extracellular substances are proteins or polysaccharides, that is, macromolecules formed by the linking together of smaller repeating molecular units. They are formed, therefore, of the same smaller units made use of within the cell, that is, starch within the cell and cellulose without are both derived from glucose molecules. Often they combine with lipids and minerals to produce highly complex molecular structures. The functions these substances perform are varied and often multiple in nature: *water retention* in the case of the slimy secretions of many algae (agar is a commercially useful product derived from some algae); *protection* as provided by the tough, chitinous covering of insects; *support* as from the cellulose walls of plants, and from the *collagen* in cartilage and bone; *rigidity* and *hardness* as from the mineralized regions of bone, the dentine and enamel of teeth, the siliceous shells of diatoms, and chitin; *elasticity* as from the *elastin fibers* of skin or artery walls; and *adhesiveness* as from the *middle lamellae* of plant cells and the *hyaluronic acid* and *chondroitin sulfate* of animal cells. The slime secreted by amebae and other aquatic organisms can provide adhesion as well as act as a lubricating material for gliding along surfaces, while the outer character of bacterial cells determines, among other things, their immunological and virulent properties. The adhesiveness of cell surfaces is particularly critical, because without it cells would fall apart after division, and multicellularity would be impossible. The *permeability* of cells is generally not affected by the presence of extracellular substances except in some plant cells where waterproof lipidlike substances such as waxes are present.

The subject of extracellular substances, therefore, is a large and varied one, but here we shall restrict our discussion to those commonly found in higher plants and animals.

PLANT CELL WALLS

Figure 5.1 illustrates, in cross section, the wall of a typical plant cell found in stems, leaves, or roots. Adjacent cells share a layer known as the *middle lamella,* or intercellular substance. It forms the first partition between two cells as they arise through cell division and acts later as an *intercellular cement* binding cells together. It appears to consist mostly of *pectin,* a substance related to cellulose since it too is formed from sugars; most of us know pectin as the additive responsible for the "setting" of jellies. There is some evidence to suggest that the middle lamella is formed by the coalescence of Golgi membranes, so some protein might also be present. The *primary wall,* lying between the middle lamella and the plasma membrane, is a secretion product of the cytoplasm. While a cell is enlarging, this wall is thin, elastic, and capable of great extension. It consists principally of celluloses and sugars of various kinds, although some proteins may also be present. The primary wall thickens only after the cell has ceased to enlarge. When growth of a cell ceases, a *secondary wall* forms between the primary wall and the plasma membrane. It

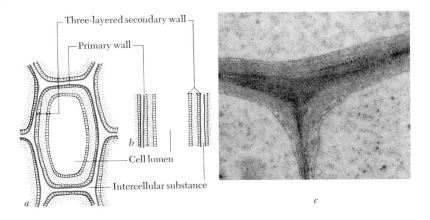

Figure 5.1 *Typical wall structure of matured and lignified plant cells.* *(a) Cross section, showing arrangement of the various layers and the complex structure of the secondary wall. (b) Longitudinal section through a similar cell. (c) Electron micrograph of the cell walls of three adjacent cells: the darker middle area is the middle lamella; the lighter portions, the primary wall.* [Reprinted with permission from K. Esau, *Plant Anatomy* (New York: John Wiley & Sons, Inc., 1953).]

may be thick or thin and of varying degrees of hardness or color. It is the part of the cell that gives various woods and plant fibers (cotton, flax, hemp) their particular character, and from which is derived the cellulose used in the manufacture of rayon, nitrocellulose, cellophane, and certain plastics.

Let us examine the growth of the cotton fiber in order to illustrate the principles of cell-wall formation. The mature fiber, or *lint* as it is called, may be 0.5 to 1.5 in. long. Located in the outermost layer of cells, or *epidermis*, of the seed coat (Figure 5.2), each lint cell is attached to neighboring cells by a middle lamella and possesses a thin primary wall. On the day of flowering and after fertilization of the egg has taken place, the cell begins to elongate, a process that takes 13 to 20 days, and terminates when the cell is 1,000 to 3,000 times as long as it is wide. The primary wall then ceases to elongate, and the secondary wall forms as sugars in the cytoplasm are converted into cellulose fibrils (presumably the conversion takes place outside of the plasma membrane), and deposited on the inside of the primary wall. Deposition of cellulose continues until the fruit is mature. The cell then dies, collapses, and flattens to give the fiber used in the manufacture of cotton threads and cloths.

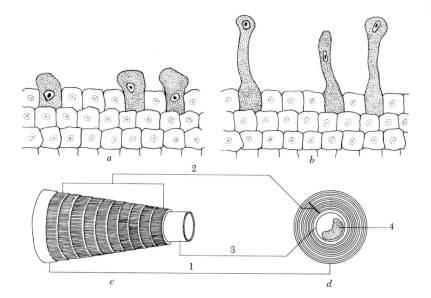

Figure 5.2 **Growth and structure of the cotton fiber.** *(a) Outer layer of cells of young cotton seed, showing the beginning enlargement of the fibers on the day of flowering. (b) Same, 24 hr later. (c) Diagram of the various layers of cellulose laid down in a mature cotton fiber: (1) outer primary cell wall, (2) concentric inner layers, revealing the different orientation of the cellulose in the secondary thickenings, (3) last inner layer of the secondary wall. (d) Same, in cross section, with (4) representing the remains of cell contents.* [Redrawn and modified from H. B. Brown and J. O. Ware, *Cotton*, 3rd ed (New York: McGraw-Hill Book Company, 1958).]

Figure 5.3 **Fibers of cellulose as formed in the wall of an algal cell.** *Each fiber would be composed of many small fibrils grouped as in a rope or cable (× 16,700).*

Cellulose is a *macromolecule* built up of repeating units of sugar (glucose) into *microfibrils* that eventually reach sizes which are relatively large in the world of electron microscopy (Figure 5.3). Cellulose is clearly formed by the activities of the cytoplasm of the cell, but how and where the fibrils are assembled and arranged in the secondary wall is not understood. In cross section, the cotton fiber has an area of about 300 μ^2 and is made up of approximately one billion cellulose chains. These are grouped into *fibrils* of several orders of size, each one running the length of the entire fiber in a parallel, or sometimes helical (spiral), fashion. The spaces between the fibrils give the fiber its flexibility and allow for the complete penetration of dyes, whereas the parallel orientation of the fibrils accounts for its great tensile strength (nearly that of steel). It has been estimated that each cotton fiber contains about 10 trillion cellulose molecules, which are built up from approximately 60 quadrillion glucose molecules. Also, a single fiber is but one of many thousands growing on the surface of each cotton seed. From these rough calculations, we can gain some appreciation of the activity of plant cells in transforming carbon dioxide and water through photosynthesis into organic molecules; multiplying these molecules, the cell then builds them into an elaborate structure, the cell wall. The cell carries out the process of repeating molecules to form its structural elements—proteins, fats, nucleic acids, and polysaccharides—in much the same way man does to form plastics and synthetic fibers, although the chemical reactions are probably different.

It appears that the arrangement of cellulose fibers initially laid down determines the kind of expansion a plant cell will undergo. When the fibers are randomly arranged, the cell expands uniformly in all directions. If the fibers are arranged in parallel fashion, expansion is at right angles to them, to give an elongated cell (Figure 5.4). The cotton fiber is elongated and its fibers parallel, or nearly so, with its long axis, because they are laid down after elongation.

The organization of the cell wall also illustrates two sound construction principles. The strength of the cotton fiber, composed of pure cellulose, is gained by grouping the molecules into ever larger fibrils; this arrangement of parts is the principle used in the construction of cables and ropes. Other types of cell walls, however, are impregnated with different substances. One of these is *lignin*, a complex, nonfibrous molecule, unrelated to the sugars, that forms in the spaces between the cellulose fibrils. This arrangement is also the principle of the reinforced concrete that is utilized in many of our modern buildings; the cellulose provides rods of high tensile strength and the lignin is a hard substance that is resistant to pressures, and acts as a glue or cement. When lignin is absent, as in balsa wood, the material is both soft and brittle. The cellulose does not have to be as well oriented as in the cotton fiber, and other substances, such as *cutin* and waxes, both derivatives of fatty acids, may replace

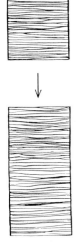
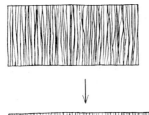

Figure 5.4 Schematic representation showing how the orientation of cellulose fibers in a young cell (top row) determines the axis of elongation of older cells. Elongation is essentially at right angles to the direction of the fibers. When the fibers exhibit no orientation (top right), the cell enlarges in all directions.

EXTRACELLULAR SUBSTANCES

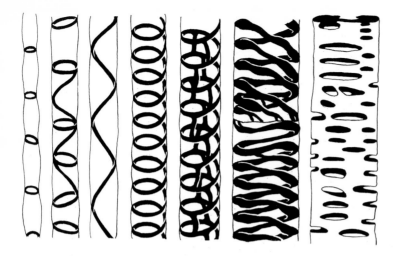

Figure 5.5 **Cells in the wood of higher plants,** *exhibiting various patterns of secondary-wall formation. The interrupted areas are thin enough to permit the passage of water and dissolved materials.*

Figure 5.6 **Water-conducting cells in the xylem of higher plants,** *showing different arrangements of pits on their side and end walls: (a) from sequoia, (b) from bracken fern, (c) from alder.*

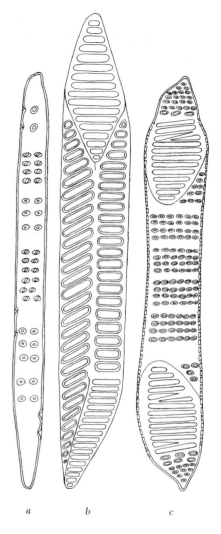

a   *b*   *c*

lignin. In such instances, the strength of the wall is less, but the cell surface is water-repellent. *Chitin,* a polysaccharide found in the exoskeletons of insects, is also present in the cell walls of many fungi, while the cell walls of grasses and particularly the horsetails contain substantial amounts of *silica,* a major component of glass and sand.

Since most plant cells conduct water and dissolved substances as well as provide support, even after they have died (as in wood), the heavy wall must be interrupted at intervals to allow for passage from one cell to another. Interruption occurs in a variety of ways, and as

Figure 5.5 indicates, the secondary wall may exist as rings, spiral bands, or sequences of thick and thin areas. Such gaps give flexibility as well as support and ease of conduction. In other cells, particular areas may be perforated by *pits*, or *pit fields*, or the end wall of a cell may be perforated or even missing to provide a connected column simulating a channel made up of short pieces of pipe (Figure 5.6).

The intercellular "glue" of animal cells is principally of two kinds: *hyaluronic acid* and *chondroitin sulfate*. Collectively, these are known as *ground substances*.

Combined with protein and lacking sulfur, hyaluronic acid is a jellylike, amorphous, viscous polysaccharide of high molecular weight. It is able to retain water tenaciously. Functionally, hyaluronic acid serves a number of purposes. As a "glue" it binds cells together at the same time that it permits flexibility; in the fluids of joints it acts as a lubricant and, possibly, as a shock absorber; in the fluids of the eye it acts to retain water and keep the shape of the eye fixed.

The viscosity of hyaluronic acid is determined, in part, by the amount of calcium present; for example, the cells of young embryos of sea urchins tend to fall away from each other if kept in a calcium-free medium, and developmental processes cannot proceed normally. Another point to note about hyaluronic acid is that it can be dissolved by an enzyme, *hyaluronidase*, present in sperm cells, and readily formed by some bacteria. The action of the enzyme in sperm cells permits sperm to penetrate the jelly coat that surrounds an egg; without it the sperm could not bring about fertilization. The ability of some bacteria to manufacture the enzyme means that they can effectively invade tissues and so spread an infection from the initial point of entry.

Chondroitin sulfate is a firmer gel than hyaluronic acid, and, like the latter, it is a polysaccharide combined with proteins. It is particularly evident in *cartilage*, where it is associated with fibrous elements such as *collagen*, and where it provides a matrix in which the cartilage-forming cells are nestled (Figure 5.7). The arrangement gives good support and adhesiveness while preserving a measure of flexibility (ears, nose, ends of ribs, new bones, respiratory tract, joints, and so on). Bone, in fact, appears first as a skeleton of cartilage, after which calcification of the intercellular substance takes place as one of the major processes of hardening.

The *basal lamina* is a somewhat more definitely organized and

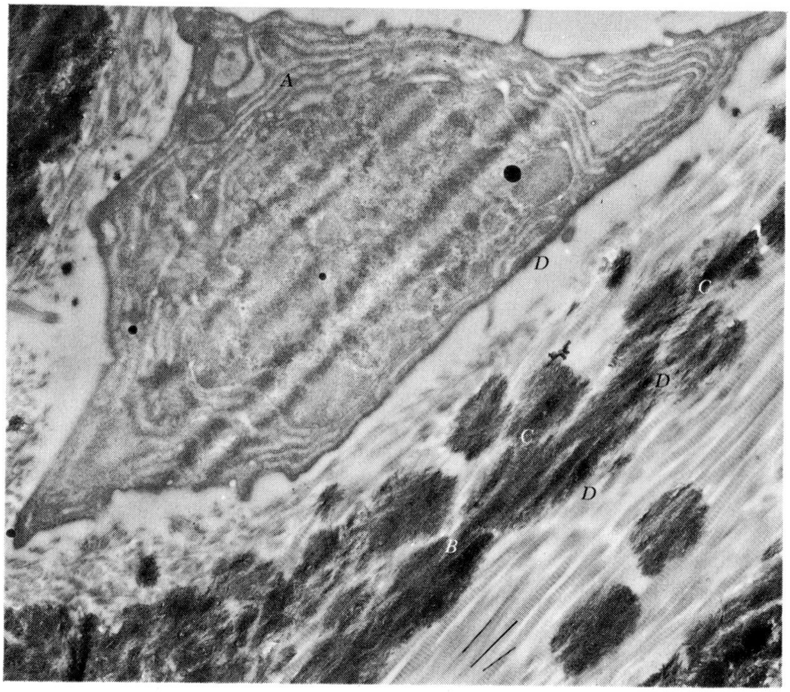

Figure 5.7 Electron micrograph of an osteoblast, or bone-forming cell, and the surrounding extracellular material: (a) Osteoblast with a large central nucleus and a well-developed ER, (b) collagen fibers, (c) collagen fibers, showing their characteristic axial periodicity every 640 Å, (d) collagen fibers beginning to decalcify, with the dark material being aggregates of individual apatite crystals of bone. (Courtesy of Dr. Melvin Glimcher.)

special ground substance. Found in a number of organs where it both binds cells together and helps to shape the particular organ, the basal lamina is prominent in skin. Here it lies between the epidermis and the dermis as a condensation of intercellular substances. Unlike ground substances in general, it may be highly laminated, as in amphibians (Figure 5.8). There may be 20 or more laminae, with fibers in each lamina lying at right angles to the layers above and below it. Such an arrangement of fibers gives flexibility as well as strength.

The fibrous elements embedded in the ground substance are *collagen, elastin,* and *reticulin.* All are basically proteins of high molecular weight. It has been estimated that about one-third of the protein of a mammal is collagen (Figures 1.4 and 5.8), and it is located in those areas where a degree of firmness or rigidity is needed (muscles, bone, skin, tendons, and so forth). Collagen can be readily dissolved in dilute acid. Dissolved collagen will reaggregate spontaneously in solution if conditions are suitable; presumably this is the manner of their aggregation outside of the cell in the ground substance.

Elastin and reticulin, unlike collagen, show no periodic structure.

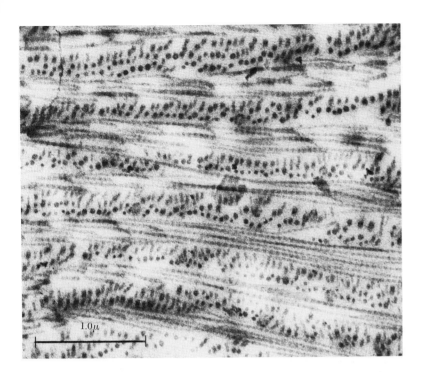

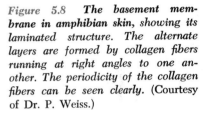

Figure 5.8 **The basement membrane in amphibian skin,** *showing its laminated structure. The alternate layers are formed by collagen fibers running at right angles to one another. The periodicity of the collagen fibers can be seen clearly.* (Courtesy of Dr. P. Weiss.)

Elastin is a stringy protein that has the capacity, as its name suggests, to stretch and snap back into its original state, much as an elastic band would. Consequently, it is prevalent where elasticity is required, as in skin and the tissues surrounding the major blood vessels.

Reticulin consists of much finer fibers than do collagen and elastin, but is probably closely related to collagen in all ways except aggregation. The fibers are finely branched, are found generally in the ground substance, and are particularly abundant in the basal lamina.

From what has been said it should be clear that ground substances and their associated fibers serve a number of purposes. Cells aggregate into organs of definite shape and size, and organ systems are tied together to form the intact and functioning organism. Adhesiveness, lubrication, rigidity, and elasticity are but some of the required features of a functioning organism that are governed by the quality and quantity of ground substance. If the character of the intercellular products changes, the organism itself must similarly change.

EXTRACELLULAR SUBSTANCES

It now seems apparent that the process of aging is, in part, related to changes in the intercellular substances. All of us are familiar with some of these changes in a gross way: stiffening of joints, loss of elasticity in the skin, hardening of arteries, toughness and stringiness of a piece of old beefsteak. There is still uncertainty about the meaning of many of the alterations in ground substance that accompany aging, but it is at least clear that an increase in the amount and thickness of the collagen fibers occurs; that the elastic fibers become thicker and less springy, possibly because of an increased binding of calcium ions; and that the reticular fibers become heavier and more brushlike. If we consider that these fibers, so necessary to the organization of living systems during the course of development, continue to be formed after mature growth has been reached, then we can view aging, at least in part, as an aspect of development that has gone beyond the optimal needs of the organism for proper functioning. Acquiring multicellularity through the adhesiveness and preservation of cells means that aging is simultaneously introduced into the life history as an inevitable consequence wherever a determinative, or limited, type of growth is characteristic—as it is in animals. Single-celled organisms are potentially immortal, and the relatively long life of plants such as a tree results not from the preservation of cells but rather from the discarding of old cells—to become wood, bark, and dead leaves—and the continued production of new cells. Plants, therefore, possess a more indeterminative type of growth than do animals.

EVEN A CASUAL ACQUAINTANCE WITH THE LIVING WORLD LEAVES ONE with the impression that there exists an almost bewildering array of forms. These vary in size from the microdomain of molecules of biological significance to the macrodomain of populations of individuals, but recognition of form, whether it be that of a molecule, cell, or individual human being, is itself recognition of the existence of some kind of order. And the more complex a form, the higher the degree of order necessary to achieve and maintain that form.

Form is a structural aspect, but living things also manifest an ordered behavior, and this derives from function as well as from form. In fact, life at all levels may be viewed as a process that organizes matter and energy into ordered systems; it is, in another sense, a method for overcoming *entropy*, the degree of randomness in a system. Order, then, must exist at all levels. This can be readily observed at the level of cells, for as discussed earlier the cell is a highly ordered and organized system compared to the environment in which it exists. As a relatively self-sustained system, the cell possesses an order that is planned rather than haphazard. This order has its origin in the

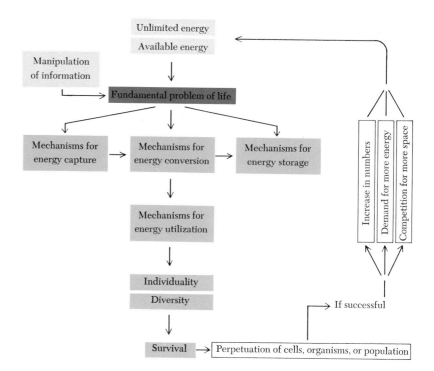

Figure 6.1 **Dynamics of living systems viewed as an energy problem, with the manipulation of information governing the movement of energy through ordered systems. (After Dr. H. H. Hagerman.)**

coded information of its DNA, and it achieves expression via the structures and reactions of the cell as it exists in a given environment. Furthermore, we recognize that the attainment and maintenance of order, whether of form or function, requires the ordered acquisition, conversion, and expenditure of energy (Figure 6.1).

Since cells, even within the same organism, may differ from each other in form and function, we should then expect that these differences would be reflected in their morphology and their management of energy at the same time that their common features are reflected in their commonly shared attributes. Here we shall concentrate primarily on the structural differences between cells, and consider only briefly their physiology and biochemistry.

Among unicellular life, form is the principal means of classification although bacteria may be grouped according to their stainability or by-products of metabolism. The protozoa and unicellular algae are

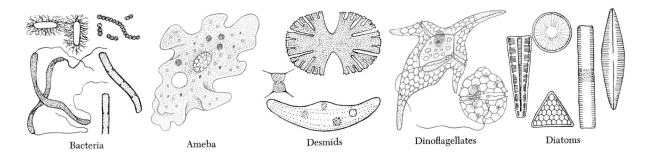

| Bacteria | Ameba | Desmids | Dinoflagellates | Diatoms |

Figure 6.2   *Examples of cell shape among unicellular organisms.*

(see also Figure 2.5)

Figure 6.3   **The human egg (a) and sperm (b), the former thousands of times larger than the latter.**

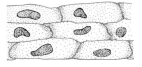

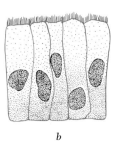

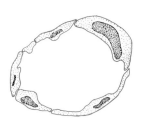

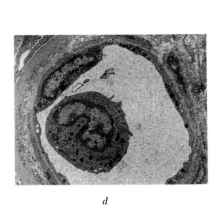

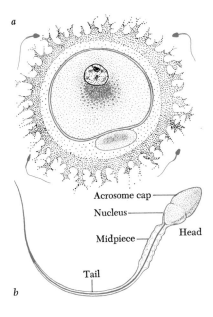

a

b

a

b

c

d

Acrosome cap

Nucleus

Midpiece

Head

Tail

Figure 6.4   **Cells of different shapes.** *(a) Cells of the skin are longer and wider than they are deep. (b) Cells of the digestive tract are columnar and far deeper than they are wide (see also Figure 2.5). (c, d) Cells lining an artery are flat and thin and are capable of expansion and contraction as the artery expands and contracts.*

COMPARATIVE CYTOLOGY

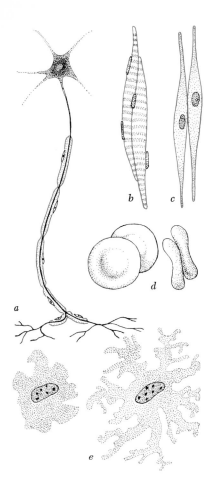

especially rich in their variety of shapes (Figure 6.2). We might ordinarily assume that a single cell, existing in a liquid environment, would possess a somewhat spherical form because this would be the simplest shape for an equitable distribution of surface tensions, that is, much as a soap bubble is rounded when free-floating. Many cells of this sort are spherical, for example, eggs of many organisms, some algae, and many bacteria. Others, however, are not, and many of these possess an extracellular surface of distinctive shape and character that is genetically determined. The diatoms and dinoflagellates among the algae have hard, silicious coverings which are fixed in shape, while many protozoa possess a thick pellicle variously smooth, ciliated, or flagellated.

Among the cells of multicellular forms, a variety of shapes are present in the same organism. Here we must assume that shape is related to function and position within the body since all cells would have a comparable amount and kind of DNA. In the human, the egg and sperm provide the most sharply contrasting forms: the egg spherical, the sperm elongate, flagellated, and distinctively organized (Figure 6.3). The shape of each is appropriate to its function and activity.

Figure 6.4 illustrates a number of other cells in the human body. Those of the outer skin tend to be longer and wider than they are deep; they form an impervious layer over the surface of the body. Cells lining the arteries and veins are somewhat similar in shape, and would be able to maintain the longitudinal shape of these organs at the same time that they can be stretched or contracted as the blood flow varies. The cells lining the digestive tract vary; the absorptive cells of the intestine are columnar, with the absorptive surface covered with tiny projections, or microvilli (Figure 2.5). Nerve and muscle cells are strikingly elongate in appearance, while those of the blood are rounded or ameboid (Figure 6.5). It is obvious that the transmission of nerve impulses or the contraction of muscles is more easily accomplished by elongated cells than either function would be by cells having the shape of a red blood cell.

The cells of a higher plant, for example, those of a flowering species, also vary according to their position and their function. Where the body of the plant is elongate, as in the stem, so too are the cells, although conducting cells are more so than the protective cells of the outer portions (Figure 5.6). Those of a leaf possess other shapes, consistent with their function, while dividing and storage tissues, as in an apical growing region of a stem or in the pith of an elderberry stem, are somewhat isodiametric but faceted by the pressure of adjoining cells (Figure 6.6).

Figure 6.5   *Animals cells of different shapes. (a) Nerve cell, with axon (long branch) and dendrites (the finer, small branches); the axon is encompassed by Schwann cells (see Figure 2.6). (b) Striated muscle cell. (c) Smooth muscle cell. (d) Human red blood cells, front and in cross section. (e) Pigment-containing melanocyte, expanded and contracted.*

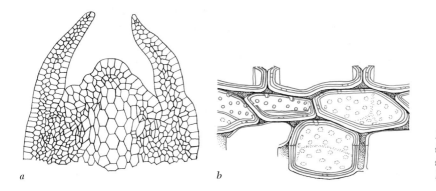

Figure 6.6 **Isodiametric cells, found in the growing tips of plants (a) and in the pith (storage tissue) of an elderberry stem (b).**

Just as cell shape varies both within and between organisms, so too does cell size. The human egg (diameter of 0.1 mm), for example, has a volume over a million times that of a human sperm, while an ostrich egg differs in volume from a pneumococcus bacterium by a factor of approximately 10^{16} or 10^{17} (Figure 6.7). Yet all these structures are intact and functional cells.

The smallest cells visible under the light microscope are the bacteria, some of whose dimensions are at the lower limits of visibility (between 0.2 and 0.3 μ). They are not the smallest cells, however. This category includes a group of organisms called PPLO's, or pleuropneumonia-like organisms (Figure 6.8). They are somewhat like viruses in size, but their organization and behavior is bacterial: outer cell membrane, ribosomes, hereditary material lying free in the cytoplasm, and a complete metabolic system of enzymes (about 40 are known) which permits them to be grown in a test tube independent of other cells.

In the human body, nerve cells are the largest, at least in longitudinal dimensions; their neuronal extensions may be a yard or more in length. Cells from striated muscle are also large, but those of the skin, liver, kidney and intestine, for example, probably average 30 μ in diameter. The *small leucocyte* (white cell) of the blood is at the other end of the size scale, having a diameter of 3 to 4 μ.

The size of a cell determines the amount of surface it can present to the surrounding environment. Metabolic reactions occur throughout the cytoplasmic mass, and the amount of cell surface governs the amount as well as the kind of materials that pass into or out of the cell. If the cell is too large, the interior may be starved for nutrients or be otherwise inefficient, but limitations of this sort can be overcome in a variety of ways. The columnar epithelial cells of the gut

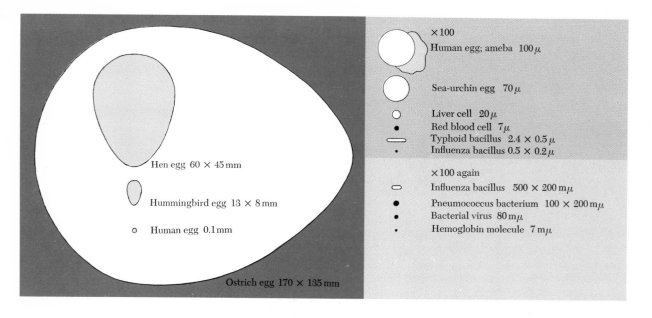

Figure 6.7 A scale of sizes of different kinds of cell, with the bacterial virus and the hemoglobin molecule included for comparative purposes. The ostrich egg and the avian eggs within it are here reduced by one-half.

Figure 6.8 Schematic drawing of the morphology of a pleuropneumonia-like organism (PPLO), with outer plasma membrane, a double helix of chromatin material, ribosomes (larger spheres), and other dissolved substances (smaller spheres). The diameter of this cell is about 0.1 µ.

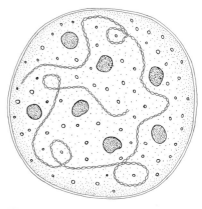

(Figure 2.5), for example, enormously increase their absorptive area through the formation of microvilli, while kidney tubule cells, governing the passage of liquids or solutes in or out of the secreted urine, have the basal cell membrane folded in a highly complex manner (Figure 6.9). Plant cells solve this problem in another way: through the formation of large vacuoles the cytoplasm is forced to the exterior where a ready exchange of gases, liquids, or nutrients can take place (Figure 3.20).

Cell size can also be viewed from the point of view of metabolic control and rate. We have discussed the fact that the nucleus is the controlling center of the cell. Like the cell membrane, the nuclear membrane is a barrier to the free passage of materials, but while it can be folded or lobed to increase the surface without alteration of nuclear volume, the nucleus itself can probably exercise control over only a limited amount of cytoplasm. A high rate of metabolism, for example, requires not only a constant supply of energy-yielding foodstuffs but also a continuing supply of ribosomes, tRNA, and enzymes which are nucleus-derived or -determined. The more rapid the rate

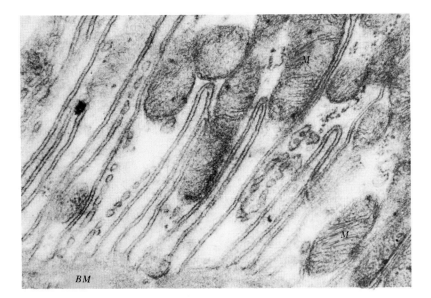

Figure 6.9 Electron micrograph of the basal (inner) side of an epithelial cell of a kidney tubule, showing the deep involutions of the plasma membrane and the numerous, closely associated, mitochondria (M). The basement membrane (BM) is indicated. The opposite end of this cell, facing the lumen of the kidney tubule, would possess many microvilli. (Courtesy of Dr. D. Fawcett.)

of metabolism, therefore, the smaller the cell is likely to be, although the correlation is not an exact one. In the animal kingdom, hummingbirds, shrews, flies, and bees metabolize at far higher rates than such organisms as man, elephants, amphibia and grasshoppers. They also have smaller cells, which are necessary in order that a proper nucleocytoplasmic ratio be maintained. In fact, if man metabolized at the same rate as a hummingbird, the heat resulting from the chemical reactions would not be able to escape either the cells or the body, and he would literally roast himself to death.

Optimum cell size is, therefore, determined by a variety of factors, all of which set upper limits to size. Lower limits must exceed a 200-Å diameter, otherwise there would be no space between the cell membranes for the cytoplasm. But the cell is an amazingly flexible unit, and the problems of size in relation to function and space have been met in a great variety of ways.

If we assume, as we have been doing, that the various structural features of the cell are related to the functions that these features perform in the general economy of the cell, then we would expect each kind of cell to reflect its performance in its particular structure. In considering a variety of cells, we have been dealing with compara-

RELATION OF STRUCTURAL
DIFFERENCES TO FUNCTION

COMPARATIVE CYTOLOGY

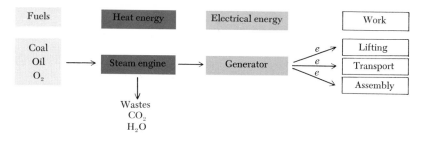

Figure 6.10   Schematic representation of the flow of energy in a typical factory, from its bound form as fuel to its final utilization to do work. The letter e represents electrical energy in the form of electrons flowing along a wire. Compare with Figure 6.11.

tive cytology in an oblique manner; here we shall confront the subject directly by comparing selected cells within a given organism—a vertebrate—to show how the special functions relate to the development of one or more structural features.

Before dealing with comparisons, however, we need to focus our attention on the central fact that *all cells are instruments of energy transformation* (Figure 6.1), and that structural differences are closely correlated with the ways energy is obtained and the use to which energy is put. The chemical details of energy relationships are the subject of another volume in this series,* so our discussion here will be in general terms.

Earlier we referred to the cell as a factory. And we know that this factory may have many sizes and shapes, and may perform either general or specialized functions. Its internal structure must be consistent with the tasks demanded of it. The energy problems, however, are basically similar in most cells.

Figure 6.10 is a schematic representation of simple factory operations in terms of energy input and output. The fuel is generally coal or oil, its energy locked up in chemical bonds. During combustion, O_2 is utilized and the energy is released as heat. The heat, in turn, makes steam to drive a generator, thereby converting heat energy into electrical energy. The waste products are CO_2 and H_2O. Electricity, in the form of electrons flowing through wires, performs work of three basic kinds: *mechanical work,* or lifting; *transport,* or moving materials from one place to another; and *assembly,* or manufacture.

The cell is basically similar in operation, but with one major exception: it cannot use heat to do work since it has no mechanism

* W. D. McElroy, *Cell Physiology and Biochemistry* (2nd ed.) (Englewood Cliffs, N.J.: Prentice-Hall, Inc., 1964).

for doing so. The primary source of energy is the sun; chloroplasts capture light energy and transform it into chemical energy in the form of carbohydrates. All other energy resides in the chemical bonds of molecules, and cells conduct their energy transactions by changing molecules.

Figure 6.11 depicts the basic steps in the energy cycle of the cell. The fuels are mainly carbohydrates, but fats and proteins may also be used. In the presence of O_2 these are broken down into smaller units in the lysosomes and mitochondria. As in our factory, the principal wastes are CO_2 and H_2O, but in the cell the chemical energy of the fuels is largely repackaged as chemical energy, and is not generally dissipated as heat. As a *charged form of energy*, the molecule ATP can yield its energy to other parts of the cell and work can be performed. The tasks again are comparable in nature to those performed by our hypothetical factory: mechanical work, as when muscles contract; transport, as when substances are moved from one area to another, often across membranes and against concentration gradients; and assembly, or biosynthesis, of other molecules and cellular structures. Energy is used to do this work, and in the process ATP is degraded into ADP (adenosine diphosphate) and inorganic phosphate (P_i). ADP and P_i are *spent forms of energy;* to be recharged, they pass back into the mitochondria where they emerge once again as ATP. The cycle must go on constantly in the living cell, and it has been estimated that a molecule of ATP in an active cell is recycled every 50 sec.

The cells we wish now to consider are those capable of performing specifically the several kinds of work that we have depicted in Figure 6.11.

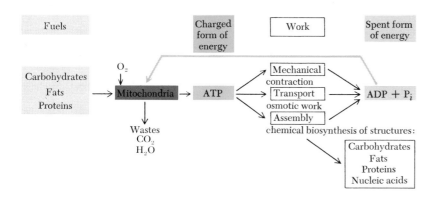

Figure 6.11   Schematic representation of the flow of energy in a cell, from the combustible fuels to its ultimate utilization for specific kinds of work. ATP and ADP + P_i are, respectively, the charged and spent forms of energy.

Mechanical work, or contraction, can be either sudden and violent or more gentle and continuous: the throwing of a baseball as opposed to rhythmic motion of breathing; the wink of an eyelid as opposed to the slow contractions of the stomach and intestines during digestion. Two kinds of cells perform these two kinds of contraction.

Striated, or *striped*, muscle cells are capable of sudden contraction; like a muscle itself, the cells are elongated and tapered, and are multinucleate (Figure 6.5). The individual cells may be from 1 to 40 mm in length, and from 10 to 40 μ in width; their several nuclei, found at the outer edge of a cell, are apparently needed to control this large mass of highly organized cytoplasm. Each cell is covered with a thin, tough membrane, the *sarcolemma*. Internally, the cytoplasm consists primarily of longitudinal filaments, called *myofilaments*, arranged systematically; they appear under an electron

Figure 6.12   Electron micrographs of portions of a striated muscle cell. (a) sarcomere, that is, the region between the two Z bands (the thin, dark bands at left and right), with mitochondria between adjacent sarcomeres. (b) Cross section through a sarcomere, showing an end view of the filaments and their definite arrangement; the larger, darker ones are myosin, and each is surrounded by six lighter strands of actin.

a

b

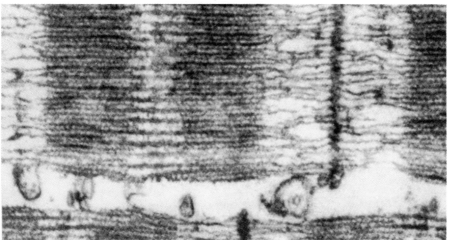
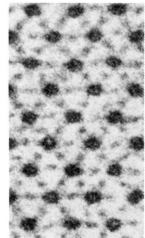

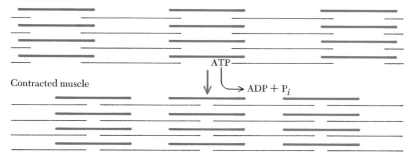

Relaxed muscle

Contracted muscle

ATP
ADP + P$_i$

Figure 6.13 **Diagram of how a muscle is thought to contract. With the release of energy from ATP, cross links are formed between actin (thin lines) and myosin (thick lines), and the actin contracts, drawing the myosin units closer together.**

microscope as alternating light and dark areas (Figure 6.12). Different bands are clearly distinguishable, and one block of myofilaments between two Z bands is known as a *sarcomere*.

During contraction the sarcomeres become shorter and broader, thereby shortening the cell. The united action of many cells, coordinated by nerve impulses, causes the whole muscle to contract, and thus perform work. The myofibrils are the agents of contraction. When viewed in both longitudinal and cross section, the fibrils are alternately thick and thin. Both types are protein, the former being *myosin* (about 100 Å in diameter), the latter, *actin* (about 30 Å in diameter). The fibrils are believed to oscillate back and forth rather constantly along the long axis of the cell, but that movement can be counteracted by the formation of chemical cross links between the two kinds of fibrils. When cross links form between myosin and actin, the thin ones contract, and thus shorten both fibers and cell (Figure 6.13). Only the dual protein molecule, called *actinomyosin*, is capable of contraction, and ATP is also needed to convert actin from a globular to a filamentous form before it can interact with myosin. The number of cross links at any given time is related to the amount of additional ATP present. Thus, when the amount of ATP is high, and the energy available is within the ATP molecule, the cross links are few in number; when ATP is split to ADP and P$_i$ and energy is released, the amount of linkage is high and contraction takes place. The following relationship, greatly simplified, thereby holds:

(1) Energy spent
(2) Cross links formed between actin and myosin
(3) Contraction of actin

ATP $\xrightarrow{}$ ADP + P$_i$

(1) Energy charged
(2) Cross links broken between actin and myosin
(3) Relaxation of actin

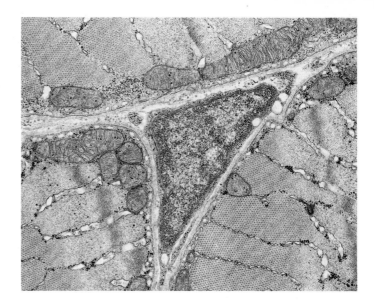

Figure 6.14   Electron micrograph of a cross section of a muscle, showing the sarcomeres cut end-on and their association with large and complex mitochondria. (Courtesy of Dr. C. W. Philpott.)

The actual details of how these actions are accomplished is still not entirely clear, but it appears that most types of cellular movement—beating of cilia and flagella, movements of microvilli and amebae—occur in a comparable manner, that is, by contractile proteins. The molecules of ATP must, of course, be plentiful and ready at hand when needed. The numerous and large mitochondria found adjacent to the sarcomeres (Figure 6.14) represent ready sources of energy, and fit the muscle cell for its task.

Smooth muscles, in contrast, are found, for example, in the walls of arteries, veins, uterus, and digestive tract (Figure 6.5); and they contract involuntarily and rhythmically. Individual cells are smaller than those of striated muscle, they are uninucleate, and they lack the tough sarcolemma of a striated cell. The striation so characteristic of the other muscle cells is also missing. Only actin is present in the form of long, thin filaments. Whether contraction occurs in a manner comparable to that described for a striated cell is debatable, but since contraction is slow and gentle, the energy requirements are less demanding. Mitochondria are consequently less conspicuous in size and number.

If we now view muscle cells in terms of work performed, we can appraise their structure more meaningfully. The mitochondria are far more numerous in striated than in smooth cells, but in each type of cell their number and size are related to the cellular demands for energy. The mature muscle cells require no substantial amount of machinery for assembly or biosynthesis of materials other than ATP, so the ER, ribosomes, and Golgi apparatus form relatively inconspicuous, but by no means dispensable, elements in the overall economy of these cells. The myofilaments, on the other hand, are our first encounter with this kind of organelle; their contractile capacity, however, is peculiarly suited to the kind of work involved, and since mechanical work (contraction) is the main function of these cells, the myofibrils loom large as a structural and functional element.

In a vertebrate there are two principal cellular transporting systems: (1) the cells of the lining of the digestive tract, which move soluble food from the lumen of the intestine into the blood stream; (2) the cells of the kidney, which extract wastes and fluid from the blood stream and excrete them as urine. The cells of both systems are comparable, at least in their grosser aspects, and we need consider only those of the kidney, since we have already depicted the columnar absorbing cell of the intestine with its numerous microvilli (Figures 2.5 and 2.7).

Transport of substances across a membrane is not a passive effort; it requires energy. We would expect therefore to find numerous mitochondria in cells engaged in active and constant transport; and indeed they are large in both number and size (Figure 6.9). Since the rate of movement of materials in and out of the cells is, in part, a function of the amount of cell surface, we would also expect to find various modifications of the plasma membrane. The microvilli on one side of the cell, and the deep indentations of the membrane at the opposite side, provide such an increase in surface area. In fact, it has been calculated that in one part of a kidney tubule, there are about 6,500 microvilli per cell, thus increasing the surface area approximately 40 times. They are similar in appearance to those of columnar epithelial cells (Figures 2.5 and 2.7).

The microvilli face the interior of the lumen of the tubule. Their function is absorption (some cells with microvilli secrete substances, however, so absorption may not be the only function of microvilli), and they reabsorb water and food materials such as dissolved sugars and salts from the urinary liquid passing down the tubule. Passing through the cell, the water and sugars leave the cell at its opposite end, through the infolded membranes at the base where capillaries abound. Water and food are thus reclaimed, but wastes pass on. In addition, however, the cells lining the tubules secrete substances into the lumen. The molecular traffic in kidney cells, therefore, is two-way. The structure of the cell is consistent with these activities, and since the degree of absorption and secretion varies along the tubule, the number of microvilli and the frequency of the deep folds in the basal membranes vary from cell to cell. Since movement of materials across these membranes requires energy, the number and size of the mitochondria also vary.

The Golgi apparatus, the ER, and ribosomes are present, but, as in the muscle cells, they are not conspicuous elements occupying large portions of the cell volume, for the materials passing through the kidney cells are neither synthesized nor repackaged in any significant way.

ASSEMBLY CELLS

A great many cells are specialized for the biosynthesis of certain substances: pancreatic cells form and secrete enzymes and digestive juices; outer skin cells, keratin; plasma cells, antibodies; and erythroblasts, hemoglobin. These substances are all proteins, and from that we know of protein synthesis, we would assume that substantial amounts of

Figure 6.15 **Electron micrograph of a plasma cell (×9,600),** *which produces antibodies; the antibodies are exported by the cell and enter the general circulatory system of the body. The inset is a higher magnification (×23,000) of the ER.* [Courtesy of A. Ham and J. Leeson, *Histology* (4th ed.), 1961. Philadelphia: J. B. Lippincott Co.]

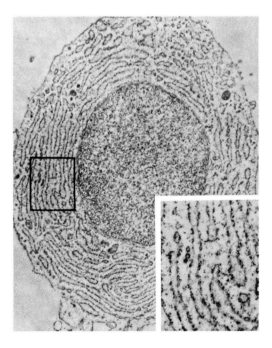

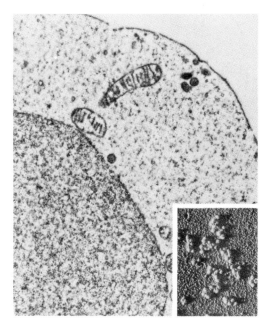

Figure 6.16   *Electron micrographs of a proerythroblast (immature red blood cell, or RBC) (×14,300) and of the polyribosomes (×36,000) (lower right) that manufacture hemoglobin. The nucleus is at the lower left.* [Courtesy of A. Ham and J. Leeson, *Histology* (4th ed.), 1961. Philadelphia: J. B. Lippincott Co.; inset, courtesy of J. R. Warner, A. Rich, and C. E. Hall.]

the several kinds of cytoplasmic RNA must be present to carry out this task. Figure 6.15 is an electromicrograph of a *plasma cell,* and the richness of the rough ER is evident. But contrast the cell in this figure, which is a secretory cell, with that in Figure 6.16, which depicts a *proerythroblast,* an immature red blood cell. It, too, synthesizes a protein, hemoglobin, but its cytoplasm is strikingly free of the membranes of the ER although rich in ribosomal particles containing RNA. The erythroblast also possesses a meager amount of Golgi membranes, whereas the plasma and pancreatic cells show a normal Golgi apparatus.

The contrasting differences here are probably related to the fate of the synthesized material. In the erythroblast, the hemoglobin remains within the cell; no further transformation or transportation of it is necessary. The plasma cells, however, and those depicted in Figures 2.12 and 2.14, prepare their protein products for delivery outside the cell; these products are apparently transformed and repackaged. From what we now know, it seems likely that the repackaging is a function of the membranous ER and the Golgi apparatus. Once again, therefore, we find that the functions performed by the cell and its internal architecture are consistent, and this consistency permits us to interpret function in terms of structure, and vice versa.

COMPARATIVE CYTOLOGY

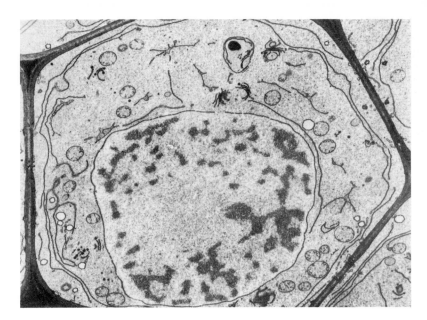

Figures 2.15 and 6.17 provide a further contrast. Figure 2.15 is a cell from the testis of an opossum, the principal function of which is the production of a steroid hormone. There is little morphological evidence of either bound or free RNA, and none would be expected in any great amount since protein synthesis is not the principal cellular activity. The smooth ER, however, is believed to contain the enzymes for steroid synthesis, and it is consequently prominently developed. Figure 6.17, on the other hand, is an unspecialized cell from a rapidly dividing tissue. Such a cell has all the organelles—the centrosome is missing since this is a plant cell—but no one element is emphasized at the expense of any other. However, since cells of this type are rapidly growing and dividing, and consequently manufacturing more nuclear and cytoplasmic materials, an abundance of ribosomal RNA in the cytoplasm is characteristic.

PROTOCELLS

THE CELL

The cells we have been considering generally are *true cells,* or *eucells.* They are distinguished by the grouping of their principal hereditary materials within a membrane-bound nucleus. Other cells, called *protocells,* lack such a nucleus, although they contain a "nuclear" substance, localized within a defined nuclear "area," and an outer, limiting plasma membrane. Whether they are more primitive than eucells, as their name would suggest, is probable although debatable, but their distinctive characteristics set them apart. Protocells are found among the bacteria and blue-green algae.

Figure 6.18 depicts a bacterial species as seen under a light microscope. Staining procedures reveal the nuclear "area," but little else in the way of internal architecture. Higher magnification by means of electron microscopy reveals additional details (Figure 6.19). The nuclear area, unbounded by a membrane, has a low electron density and is therefore relatively light in appearance, but the fine filaments are the hereditary material (DNA) of the cell. This may be seen in Figure 1.4; the bacterial cell has been burst, and its "chromatin" takes the form of a long, thin strand having a diameter of about 20 to 30 Å, as expected for a DNA double helix. The cytoplasm may contain some membranous structures, possibly comparable to the ER of higher cells, and is richly packed with ribosome particles containing RNA.

The cells of blue-green algae are somewhat more highly organized. The chromatin of the nucleus is not limited by a membrane, but numerous membranes in the cytoplasm represent the photosynthetic lamellae. However, these membranes are not grouped into distinct chloroplasts.

We can, therefore, visualize a series of cells possessing an increasingly more complex set of organelles. The bacteria are the least complex, the blue-green algae a step above them in this regard, and then the true cells of more advanced plants and animals. Whether this sequence represents the stages of evolution in the development of cells is an open question. Since the plastids of higher plants contain DNA and photosynthetic membranes (grana), the hypothesis has

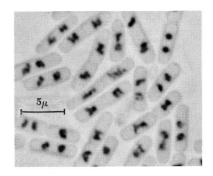

Figure 6.18 **Cells of** Bacillus cereus *stained by the Feulgen method to show the nuclear area.* (Courtesy of Dr. C. F. Robinow.)

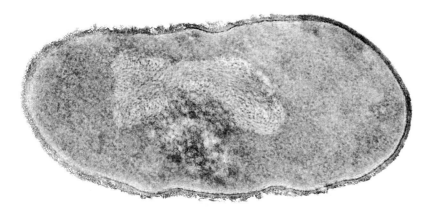

Figure 6.19 **Electron micrograph of a germinating spore of** Bacillus subtilis, *showing cell wall, darker plasma membrane, light nuclear area, and a relatively undifferentiated cytoplasm* ($\times 52,500$). (Courtesy of Dr. C. F. Robinow.)

COMPARATIVE CYTOLOGY

been advanced that plastids are derivatives of blue-green algae cells that have taken up residence in another cell, and have, during the course of evolution, become adapted to this symbiotic existence. The hypothesis is not as startling as it may seem, for plastids, under some circumstances, display a form of inheritance somewhat independent of the cell of which they are a part. Mitochondria also contain DNA, and may also exhibit "mitochondrial" inheritance, but if they are foreign organisms which have become an essential "element" of normal cells, we have, as yet, no knowledge of their evolutionary origin.

VIRUSES

While on the subject of comparative cytology, we should consider the viruses. They are living organisms, but they are not cells in the conventional sense. More than three hundred different viruses are known. Many of them are infective agents in such diseases as yellow fever, rabies, poliomyelitis, smallpox, mumps, and measles in human beings, and in a wide variety of diseases in plants and animals. Figure 6.20 shows several viruses as they appear in the electron microscope.

The metabolism and structure of viruses differ from that of cells with which we are acquainted. A virus, for example, is not a free-

Figure 6.20 **Electron micrographs of viruses.** *(a) Tobacco necrosis virus: the virus particles are spherical and about 250 Å in diameter; but when precipitated in ammonium sulfate, they form a crystalline structure. (b) Tobacco mosaic virus (TMV): each rod is made up of a stack of plates similar to a stack of coins, with a protein coat on the outside and an inner core of RNA. (c) P2 bacteriophage of bacterial virus, which attacks the colon bacterium: each is equipped with a somewhat hexagonal head and a tail. (d) T6 bacteriophage, which also attacks the colon bacterium. (Courtesy of Dr. L. W. Labaw.)*

a   b   c   d

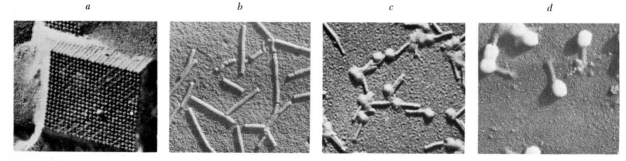

living organism; it can grow and multiply only within another cell which is has parasitized. In some manner, the virus can alter the metabolic machinery of the cell so that it is directed toward making more virus instead of performing its usual functions. In other ways, however, viruses are clothed with all the characteristics of life: they grow, multiply to produce exact replicas of themselves, and possess a type of inheritance not far different from our own. They also contain the key molecules of protein and nucleic acid invariably found in every living organism. The closest comparison that we can make between viruses and cells—if such a comparison is legitimate—is that the viruses are cells without cytoplasm; they possess little more than the nuclear hereditary apparatus and an outer limiting coat of protein, but the protein coat does not have the characteristics of a cell membrane. Probably the best guess as to their origin is that they are degenerate cells adapted exclusively to a parasitic existence.

7 THE CELL IN DIVISION

THE BODY OF A MAN OF AVERAGE HEIGHT AND WEIGHT CONSISTS OF about 10^{14} cells. Life for this man, as an individual, began when an egg was fertilized by a sperm to form a zygote. During the course of development the zygote is transformed into an embryo and then into a recognizable individual by the processes of cell division, cell differentiation, and cell death. These processes, occurring at the proper time and in a given sequence, give form to the individual and provide it with the organ systems necessary for meeting the demands of life in a given environment. Cells, therefore, do four things in addition to their metabolic functioning: they fuse, differentiate, die, and/or divide. In this chapter, we shall deal with the process of cell division.

In any multicellular organism that grows or requires repair, the process of division produces new cells at a rate and for a period of time demanded by the inherent needs of the organism and its internal and external environmental circumstances. Each kind of organism will differ in its general patterns of division. An oak tree, for example, has its main period of growth during late spring and early summer, with the dividing cells located at the tips of branches and roots and around the outside of branches and roots just under the bark. The first group of cells extends the branches and roots; the latter group contributes to its increasing girth. A mammal, on the other hand, has active centers of division in the skin, the blood-forming regions of the long bones and certain cells of the intestinal lining, slower division rates in the liver and kidney, and little or no division among nerve cells. The latter, if lost, cannot be replaced. In fact, in an adult body, most cells do not divide, although some may be induced to do so experimentally.

The process of cell division is basically the same in all organisms that have a membrane-bound nucleus. If we describe the process as it takes place in one or two kinds of cells, we can gain a reasonable idea of how it operates in virtually all other organisms. The most obvious events of cell division take place in the nucleus. The chromosomes shorten and thicken and are finally separated into daughter cells, to provide them with similar genetic constitutions. But the cytoplasm also participates in the process, and we now recognize that the entire process of division is a complex, cyclical affair, consisting of a number of sequential or parallel steps, with each one being dependent on, or conditioned by, preceding steps. Each of these steps can also be modified or inhibited experimentally, but the usual result is that the sequence is disturbed, and an abnormal division ensues. The chemical events of cell division are not so well understood as are the morphological changes, and the major emphasis here will be on those aspects that can be detected readily in cells viewed with the light microscope.

The roottips of growing plants or germinating seeds provide a ready source of dividing cells. These may be prepared for study in two different ways. By one technique, the roottip is fixed, embedded in paraffin, sectioned longitudinally or in cross section on a microtome, and then stained with some basic dye. By the second technique, the roottip is fixed, stained (generally with Feulgen, which is specific for the DNA in the chromosomes), and then "squashed" or flattened on a slide so that a monolayer of cells is formed. Figure 7.1 illustrates cells prepared by these techniques, each of which has its advantages and disadvantages.

Since cell division is a cyclical affair (Figure 7.2), a discussion of the process can begin at any point. We will start, however, with cells in *interphase* (Figure 7.3). The nucleus is readily visible, and *nucleoli* can be seen within the nucleus, but little differentiation of

Figure 7.1 *Panoramic view of sectioned and smeared roottip cells. (a) Sectioned view of dividing cells in the onion root, stained with iron hemotoxylin to show chromosomes, spindle, walls, and cytoplasm; the various stages of division range from interphase to telophase. (b) Smeared cells from the root of the broad bean, Vicia faba; smearing disrupts the arrangement of the cells, while the Feulgen stain used in this instance is specific for chromosomes and stains no other part of the cells (×410). [(a) General Biological Supply House, Inc.; (b) courtesy of Dr. T. Merz.]*

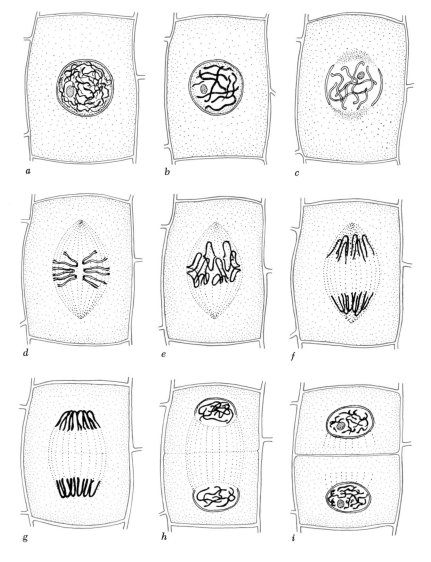

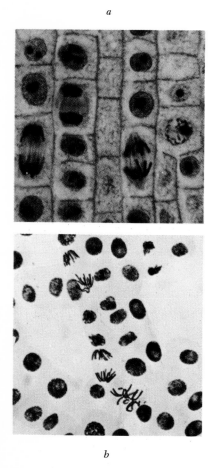

Figure 7.2 **The progress of cell division, outlined in schematic form.** *As the cell prepares to divide, the chromosomes appear as distinct bodies in the nucleus, with a split along their length. The spindle appears at metaphase and separates the two chromatids of each chromosome at anaphase, after which the cell plate cuts the cell into two new cells. Karyokinesis, or mitosis, refers to the nuclear events of cell division; cytokinesis refers to the division of the cytoplasm by the cell plate.*

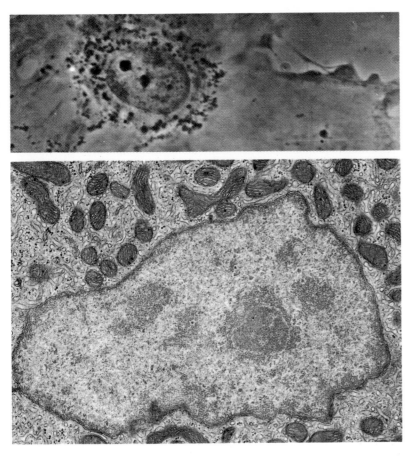

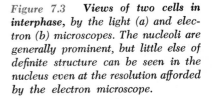

Figure 7.3 **Views of two cells in interphase, by the light (a) and electron (b) microscopes. The nucleoli are generally prominent, but little else of definite structure can be seen in the nucleus even at the resolution afforded by the electron microscope.**

b

chromatin is discernible. Both the nucleus and the cytoplasm will now undergo considerable enlargement, and nucleic acid and protein synthesis will proceed actively. The synthetic activity is indicated by the fact that if radioactive amino acids or nucleotides are given to the cell as tracers, they are incorporated readily into cellular components. The amino acids—leucine, for example—would be incorporated into both cytoplasmic and nuclear structures, thymidine would find its way exclusively into the chromatin DNA and at a specific time during the cell cycle, and uridine would be found in the newly synthesized RNA's of the cell. Interphase is, therefore, a period when

THE CELL IN DIVISION

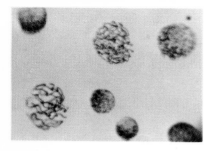

Figure 7.4 **Several stages of prophase in roottip cells of** V. faba, *with the strands of chromatin becoming increasingly visible as the cells near metaphase. Feulgen stain was used so that only the chromatin is stained.* (Courtesy of Dr. T. Merz.)

Figure 7.5 **Late-prophase stage in a spermatogonial cell of the amphibian** Amphiuma. *Each chromosome is longitudinally split into two chromatids, and the centromeres are indicated by the constricted region in each chromosome. The fuzziness of the chromosomes is due to projecting loops of fine chromatin; these would be withdrawn into the body of the chromosome by full metaphase.* (Courtesy of Dr. Grace Donnelly and Dr. A. H. Sparrow.)

the cell is preparing for division, and it generally occupies about two-thirds of the cell cycle.

The cell enters *prophase* when the chromosomes become visibly distinct as long thin threads (Figure 7.4). The nucleus resembles a loose ball of yarn, and the chromatin threads can be seen to be longitudinally double. The increasing visibility of the chromatin is owed principally to the fact that the chromatin is being compacted by a process of coiling. A long slender thread of chromatin is transformed into a compact, recognizable chromosome much in the same way that you might turn a thin wire into a coiled spring. As prophase progresses, the coils increase in diameter as they decrease in number, the chromosomes become distinct entities, and each can be seen to be made up of two longitudinal halves, or *chromatids* (Figure 7.5).

Other events are also taking place in prophase. The nucleoli, which are formed by particular chromosomes, are initially large and prominent, but they gradually diminish in size, detach themselves from the chromosomes, and usually disappear. The nuclear membrane breaks down and disappears, and the chromosomes congregate near the center of the cell.

Metaphase is now initiated, and with it, a new structure, the *spindle,* which consists of protein fibers or tubules oriented longitudinally between the *poles* (Figures 7.6 and 7.7). Chemical analysis of cells has indicated that approximately 15 percent of the cytoplasmic

Figure 7.6 **Stages in cell division in onion roottip cells:** *(a) Metaphase. (b) Early anaphase and late telophase, with cell plate forming across the spindle. (c) Late anaphase and early prophase. These cells are stained with hemotoxylin, and walls, cytoplasm, and spindle show as well as the chromosomes (×475).* (General Biological Supply House, Inc.)

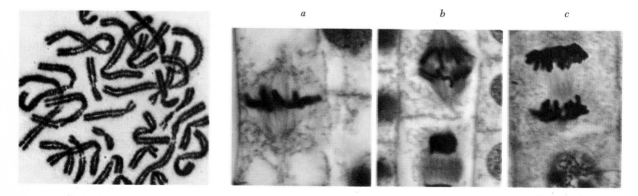

a   b   c

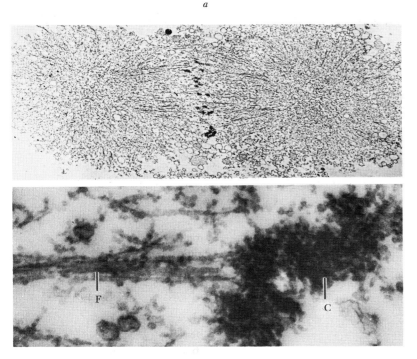

a

b

Figure 7.7 **Electron micrographs of spindles and spindle structure in sea-urchin eggs.** *(a) Isolated spindle at low magnification (×2,100), with the chromosomes appearing dark on the metaphase plate, a vague region across center of spindle; (b) spindle fibers (tubules) (F) attached to the chromosome (C) (×53,000). (Courtesy of Dr. R. E. Kane.)*

proteins go into spindle formation. Once the spindle is formed, the chromosomes become fastened to it by their *centromeres* at a region midway between the poles. This is the *metaphase plate*, which is apparently a region of equilibrium between the two poles.

It is relatively easy to prevent the spindle from forming without interfering with the preceding events of cell division. The drug *colchicine*, commonly used to relieve the pain of gout in human beings, prevents the aggregation of proteins into a spindle. The chromosomes then lie free in a cell, and their morphology can be readily determined.

Figure 7.8 shows the chromosomes of the broad bean, *Vicia faba*, as seen at metaphase. Each one has a distinct morphology that is characteristic. The location of the centromere is constant, and is identified by the constriction it forms, dividing each chromosome into two *arms* of variable length. The centromere is the structure concerned with movement of the chromosome. Without it, a chromosome cannot orient on the spindle and the chromatids cannot segregate from each other. In the longest chromosome found in the broad bean,

THE CELL IN DIVISION

Figure 7.8 **Metaphase and ana-**
phase in the V. faba **roottip. (a) Chro-**
mosomes at the metaphase stage, but
with the spindle absent as a result
of treatment with colchicine. The
centromeres are visible as constric-
tions, and the pair of chromosomes
in the upper right-hand corner show
the gap where the nucleolus was
formed. (b) Late anaphase, with the
centromeres aggregated at the poles.
The nucleolar gap is visible, particu-
larly at the left of the two groups of
chromosomes.

another constriction is also present. This chromosome formed a nucleolus at that point, and the constriction, or gap, which is an uncoiled region of the chromosome, is the site occupied by the nucleolus before it disappeared.

Anaphase follows metaphase in the mitotic cycle (Figure 7.8). The centromeres now separate so that each chromatid has its own centromere; these move apart from each other to initiate a slow movement that will take sister chromatids to opposite poles. At the end of anaphase the chromosomes form a densely packed group at each pole.

At this point, *telophase* begins. The events are essentially the reverse of those occurring in prophase: the nuclear membrane forms from remnants of the ER, the chromosomes uncoil to become a densely staining chromatin network, and the nucleoli make their appearance. A cell wall now cuts across the middle of the cell (Figure 7.9). Initially the *cell plate*, as it is called, is formed as a disc within the confines of the spindle, and apparently from elements of the Golgi apparatus or ER, but it soon increases its diameter until it reaches the walls of the cell, segmenting the cytoplasm into two roughly equal parts. The spindle then disintegrates, cell division is completed, and two new cells have been formed.

a

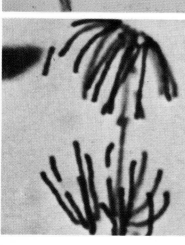

b

Figure 7.9 **Electron micrograph of a maize cell in late telophase, with the**
cell plate forming across the center. (Courtesy of Dr. G. Whaley.)

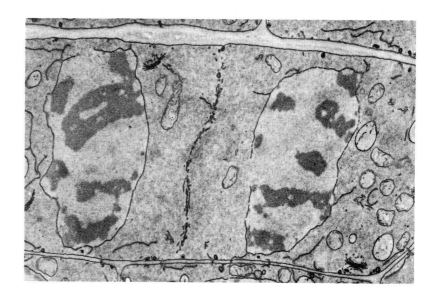

The end result of division is the same in both plant and animal cells: the formation of daughter cells of like genetic constitution. This stems from the fact that the chromosomes in each behave similarly. Differences, however, exist, and the division of cells in the embryo of the whitefish reveal these in striking fashion. The chromosomes of the whitefish are smaller and more numerous than those in the broad bean, and consequently more difficult to distinguish individually, but the most immediate difference is in the appearance of the spindle. As Figure 7.10 indicates, the spindle makes its appearance in prophase as a radiating structure adjacent to the nuclear membrane. This is the *centrosome*, with *astral rays* extending out from the center and with a *centriole* (not visible in Figure 7.10) contained within it. It is the centriole which apparently organizes the proteins into a spindle

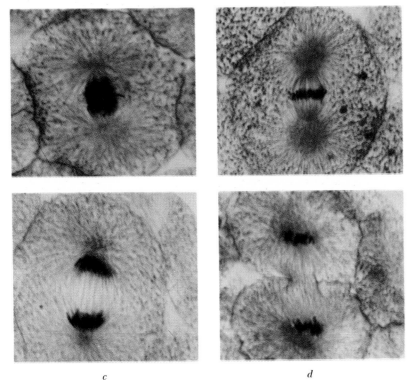

a b

c d

Figure 7.10 **Stages in division in the whitefish:** *(a) prophase, with spindle beginning to form; (b) metaphase; (c) anaphase; (d) telophase, with the furrow cutting the cell into two new daughter cells.* (General Biological Supply House, Inc.)

structure, although the mechanism of formation is not understood.

Figure 3.15 is an electron micrograph of a centriole, while Figure 3.16 shows the huge centriole found in certain protozoa. The latter figure shows that the centriole is already replicated a division in advance; the new centriole will grow to full size during the coming interphase.

In midprophase the already divided centriole separates into its two halves, and these migrate along the nuclear membrane until they lie opposite each other. As they migrate, the spindle fibers are formed between them so that when the nuclear membrane breaks down, the chromosomes are encased within the developing spindle. The position of the centrioles determines, therefore, the axis of division of a cell, governing in this way the manner in which cells are related to each other in a multicellular tissue. The two centrioles are at the poles of the spindle, and the fibers extending from pole to pole form the spindle while those extending into the cytoplasm are the astral rays. The chromosomes become attached to the spindle at the metaphase plate, and anaphase movement segregates the sister chromatid to opposite poles. No structure comparable to the centriole exists in the broad-bean roottip cells, and the manner of spindle formation is unknown although in both plant and animal cells it appears that the centromeres of chromosomes also participate in the organization and orientation of spindle fibers.

Division of a whitefish cell differs from that in a plant cell. A process of *furrowing*, beginning at the outer edges of the cell and midway between the poles, cleaves the cell in two. Plant cells, with their rigid cell walls, are not as flexible, but the process of cell plate formation accomplishes the same result.

THE SEQUENCE OF EVENTS
IN CELL DIVISION

Many aspects of cell division are not explained by a simple description of events taking place in a cell as it progresses through its cycle from one interphase to another. It must be apparent, however, that the cell cycle is not a single process, but rather is a complicated affair in which the nucleus, cytoplasm and their constituent parts come in on cue to play their necessary roles. Just as a watch runs only when all of its parts function properly, the cell completes its division to form daughter cells only when all of the elements are in a state of readiness at the proper time in the cycle, and when preceding events prepare the cell for the events to follow. For example, if the spindle is prevented from forming, the chromatids cannot segregate from each other, no anaphase movement occurs, and the cell cannot partition the cytoplasm by furrowing or cell-plate formation.

Figure 7.11 diagrams the events taking place before and during cell division. The events occurring prior to prophase, although not visible under the microscope directly, are of primary importance because they control and prepare the cell for the more dramatic events of prophase, metaphase, anaphase, and telophase. Indeed, it now appears that a cell prepares to divide long before the actual division takes place, but whether division does occur is determined, at least in part, by the metabolic activities of the cell; that is, the division process itself places a heavy metabolic demand on the cell, and if the cell's metabolism is shunted in another direction (for example, to make hemoglobin or insulin) it is less likely to be able to prepare for division. If such a cell making another product unrelated to division is to divide, its manufacturing machinery must be turned off and the divisional machinery turned on.

Centrioles replicate an entire cell cycle in advance; this act does not, therefore, determine whether a cell will divide or not. Replication of the chromosomes, however, appears to be the first crucial step.

Figure 7.11 **Table of events taking place in preparation for, and during, cell division.** (After D. Mazia.)

Preparations for division			Division	
Interphase	Prophase	Metaphase	Anaphase	Telophase
			Separation of sister chromatids	
Replication of chromosomes				
	Shortening of chromosomes			Uncoiling of chromosomes
	Synthesis and organization of spindle proteins		Movements of chromatids to poles	Disappearance of spindle
			Spindle elongation	
	Disappearance of nucleoli			Reappearance of nucleoli
		Disappearance of nuclear membrane		Reappearance of nuclear membrane
		Movement of chromosomes to metaphase plate		Division of cell
		Connection of centromeres to poles		Replication of centrioles

The cell cannot replicate its DNA and form the necessary histones of the chromosome if it is engaged in any other significant metabolic activity. The mode of replication of DNA was discussed in Chapter 4 (Figure 4.12), and the essential correctness of this concept of replication for DNA in the chromosome is indicated in Figure 1.6. When radioactive thymidine is given to dividing cells in interphase, autoradiographic techniques show the chromosomes in the following metaphase to be radioactive in both chromatids, but that in the second metaphase one chromatid is radioactive and the other is not. (Figure 4.12 diagrams these events, and serves equally well for the replication of DNA or of chromosomes.) The manner in which DNA is organized, along with histones, into the fabric of the chromosome is not yet understood, but it is clear that the replication of DNA and of the chromosome provides an explanation of how cell division leads to the formation of daughter cells of like genetic constitution. It is by this means that inheritance, embodied in the DNA molecule, is passed from cell to cell and from generation to generation.

The nucleus of a human cell contains about eight billion nucleotide pairs, distributed among the 46 chromosomes. Each chromosome contains an average of about 175 million nucleotide pairs, with one turn of the DNA double helix for every 10 nucleotide pairs. There are, therefore, about 17.5 million turns per average chromosome, and these must untangle if the chromatids are to separate properly from each other at anaphase. How this is accomplished in the cell we do not know, nor is it known if the DNA in a chromosome extends continuously from one end to the other, or whether it exists in many short segments.

Figure 7.11 also indicates that the fibrous proteins of the spindle are synthesized, at least in part, during interphase. Chemical analysis of isolated spindles indicates a single type of protein combined with about 5 percent RNA. There is also ATP present for energy purposes. With the beginning of prophase, the spindle proteins are present in the cell, but in an unassembled or unorganized form. Most of the protein appears to be of cytoplasmic rather than nuclear origin, although both parts of the cell may contribute spindle substance. Just before metaphase, the spindle proteins are oriented longitudinally between the centrioles (or the poles when no centrioles are evident), with some of the protein organized into distinct microtubules connecting centrioles and the centromeres of chromosomes. Some microtubules may extend from pole to pole, or simply project into the cytoplasm as astral rays.

The organization of spindle proteins brings the centromeres of chromosomes onto the metaphase plate, a position of equilibrium.

With the onset of anaphase, the fibers between centriole and centromere shorten, separating the chromatids and moving them closer to the poles. Additional separation of the chromatids is brought about by an elongation of that region of the spindle lying between them. The mechanisms involved in these two kinds of anaphase movement are not yet fully understood, however. After the separation, nuclei are formed at the poles, and the cytoplasm divides into two cells.

Figure 7.11 depicts still other events that transpire during cell division: the shortening of the chromosomes by coiling in prophase and their uncoiling at telophase, the disappearance and reappearance of nucleoli and nuclear membrane, the formation of a cell plate or a furrow to divide the cytoplasm, and the disappearance of the spindle. All of these, like the replication of DNA and the formation of spindle proteins, are chemical events, but although we can follow the morphological course of events during the cell cycle, we know little of the basic biochemistry of any one of these phenomena except the replication of DNA.

How long does it take a cell to go through the entire process of division and how much time is spent in each stage of division? One can determine how long it takes for a bacterial culture or a group of cells in tissue culture to double the number of cells. This would provide an average rate of cell division. Individual living cells can also be examined by phase-contrast microscopy (Figure 7.12), and the length of each stage determined. Autoradiography, making use of radioactive thymidine, can determine the length of time in interphase it takes for the chromosomes to complete their replication. These events are depicted in Figure 7.13. A human cell in tissue culture at 37°C, for example, completes its cycle in about 18 to 22 hr; a broad-bean roottip cell takes about the same amount of time at a temperature of 22°C. Yet from the beginning of prophase to the end of telophase takes less than 1 hr for a human cell, a somewhat longer time for a roottip cell. Thus the cell spends many hours preparing for division, and then goes through the visibly evident stages in rapid fashion. Interphase can also be broken down into three main subdivisions, the most easily determined one being the period of synthesis when DNA is taking up radioactive thymidine. This takes about 6 hr for both mammalian and roottip cells. The G_1 period (G stands for gap) precedes the period of synthesis (S), and G_2 follows it. We know less of what transpires in these two stages, but they are undoubtedly active metabolic stages, necessary for the completion of in-

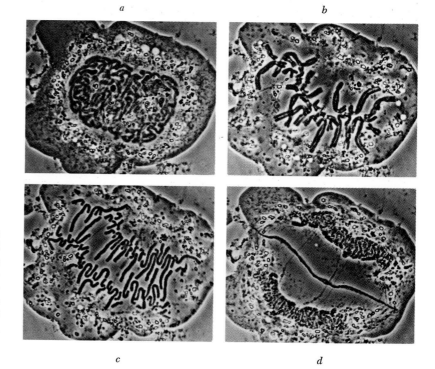

a   *b*

c   *d*

Figure 7.12 **A living cell of the blood lily,** Hemanthus, *photographed at successive stages of mitosis through a phase-contrast microscope: (a) prophase; (b) metaphase; (c) anaphase; (d) telophase. If (a) is considered to be at time 0, (b) occurs about 170 min later, (c) after another 80 min, and (d) about 90 min after (c). (Courtesy of Dr. William Jackson.)*

Figure 7.13 **An approximate division of the cell cycle into the several recognizable stages.** *For* Vicia *roottip cells and human cells in tissue culture at 37°C, the entire cycle takes about 22 hr, with the S period taking about 6 hr.*

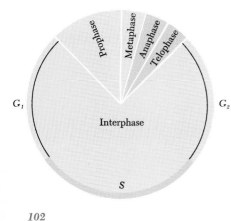

terphase and the initiation of the events to follow.

What determines the rate of cell division? Bacteria can divide in 20 to 30 min, with DNA replication being an almost continuous process through the cell cycle (bacteria show no mitotic stages and no spindle, and the mechanism of chromosome segregation differs from that of nucleated cells). A fertilized frog's egg divides five times in the first 7 hr, later embryonic cells divide about once every 24 hr, and dividing cells in the tadpole stage do so once every 4 or 5 days. Temperature can obviously regulate the rate of division of any cell—some plant cells can divide, albeit very slowly, at 0 °C—but if a nucleus is transplanted from a late embryonic stage of the frog to the fertilized egg, it will divide at the rate characteristic of the egg. The cytoplasm, therefore, can act in a regulatory manner. In fact, cells that ordinarily would not divide again can sometimes be induced to do so, but in order to do so the cytoplasmic machinery must be shifted from whatever it is doing to those activities related to cell reproduction.

Let us ask another question. Why is it that cells divide? Or, in reverse, why is it that reproducing cells stop dividing? If we take a cell such as an ameba, we can observe that it reaches a certain size and then divides. If we starve it, it will shrink and stop dividing. It would appear, therefore, that cell division is an attempt to keep a fairly constant ratio between the amount of nucleus and the amount of cytoplasm. This view makes sense, for if the nucleus governs the activities of the cell, it can exercise efficient control over only a certain amount of cytoplasmic material. If the cytoplasm should exceed a certain amount, the power of the nucleus over it becomes less and less. For example, an ameba can be prevented from dividing by cutting off bits of the growing cytoplasm each day, thus keeping the ratio of cytoplasm to nucleus constant. Yet the problem isn't quite so simple as it might appear. If the nucleus of an actively dividing ameba is inserted into an ameba that is in a quiescent state (this can be done with a micropipette without damaging either the nucleus or the cytoplasm), nothing happens unless both the nucleus and the cytoplasm are ready for cell division to occur. What we mean by a state of divisional maturity is still somewhat uncertain, but obviously both of these parts of the cell must be prepared for division before the process of division can go on in any organized fashion. Furthermore, the nucleocytoplasmic ratios may vary widely, from 1:1 in human lymphocytes to 1:1,000 in muscle cells. Yet both can be induced to divide under special circumstances. Our knowledge of the exact reasons for cell division is thus still incomplete.

Cell division is, of course, part of the process of growth. Although the dance of the chromosomes, the formation of the spindle, and the formation of daughter cells are the more obvious parts of the drama, it also involves the assimilation of materials from the outside, their transformation through breakdown and synthesis into new cellular parts, and the utilization of energy. Cell enlargement also takes place. We know of no cells, except the fertilized egg and a few of its derivative cells, that simply divide from one large cell into two others of half size and again into four cells of quarter size. This is not the usual way cell division proceeds, for interspersed in the process are periods of growth; each division is a tumultuous affair, from which the cells must recover before proceeding again through the cycle.

Of great significance to growth, too, is the fact that cell division insures a continuous succession of similarly endowed cells. Chaos would result if only a random array of cells of varying qualities and capacities were to reproduce themselves; organized growth must proceed from cells of similar nature that can subsequently be molded according to the demands of the species. The species could not otherwise persist. We mentioned earlier that the chromosome is an intricate fabric composed of nucleic acids and proteins. Since the nucleus is the control center of the cell, and since the nucleus contains little else but chromosomes, the chromosomes must be the regulators of cellular metabolism and the structural characteristics of the cell. Therefore, if two cells are to behave similarly they must have the same amount and type of nucleic acids and proteins. The longitudinal duplication of the chromosomes into identical chromatids and their segregation to the poles at anaphase must be exact to the minutest degree; the kind of cell division described provides the mechanism needed. From the time a particular species was formed, this process of cell division has gone on with an exactitude that almost defies the imagination. Accidents and variations do occur, and indeed they must if evolution is to take place, but they are relatively few in number.

Looked at in another way, cell division is an act of survival. Cells eventually die if they do not divide, just as multicellular organisms must die if they possess no way of subdividing their bodies to produce new organisms. Division, however, accompanied by the usual growth that follows division, brings fresh substance into the cell, which effectively prevents aging and gives to the cell a potential immortality. In a multicellular organism, on the other hand, cell division provides added cells, among which a division of labor can take place. Viewed in this manner, cell division is, therefore, a first step towards cell differentiation. But this is the antithesis of survival, because differentiation is also a first step toward eventual death, since differentiated cells lose their capacity to divide. The significance of cell division, then, depends not only on the phenomenon itself, but also on the kind of cell that is dividing and the consequences of division to an organism.

THE CONTINUATION OF ANY SPECIES, MAN OR AMEBA, OAK TREE OR BAC-
teria, depends on an unending succession of individuals. No organism
is immortal; so the individuals of a population must reproduce if the
species is to escape extinction. In unicellular organisms such as the
ameba, cell division, as described in the preceding chapter, serves this
function; it is a reproduction device that leads to the continued formation of new individuals. And, since mitosis is a mechanism that maintains a constant chromosome number, all offspring arising through
mitosis have the same number of chromosomes as the original ameba,
barring, of course, any accident of mitotic division.

The ameba, however, like many unicellular and some multicellular
organisms, is asexual; it does not produce sexual cells—*eggs and sperm*
—and all descendant individuals have a unilateral or uniparental inheritance. But other unicellular and most multicellular organisms reproduce by sexual means; at some time during their life cycle they produce
gametes (a general term applied to any type of sexual cell) that unite
in pairs to form a single new cell called a *zygote*. From this cell a new
individual develops, and it is the product of biparental inheritance.

The union of gametes is called *fertilization* or *syngamy*.

It is important to recognize that when two gametes unite through fertilization, the principal event is the fusion of gametic nuclei. Let us consider what this means in terms of chromosome number. The cells of the human being, for example, contain 46 chromosomes (Figure 8.1). If we assume, for the moment, that mitosis is the only type of nuclear division, the human egg and sperm would each contain 46 chromosomes since they arise by division from the original zygote. The zygote formed by their union would then contain 92 chromosomes, and so, too, would the eggs and sperm produced by the individual developing from the new zygote. The individuals of the next generation would possess 184 chromosomes, and by the end of the tenth generation each individual would have cells containing 23,332 chromosomes.

Obviously, this would be a ridiculous state of affairs. The illustration is used merely to emphasize that, in a sexually breeding population, the increase in chromosome number resulting from fertilization

Figure 8.1 **The chromosomes of a normal human male, with the chromosomes arranged in homologous pairs and numbered according to size. The male has an XY sex-determining system; the small Y chromosome is indicated at the bottom right, while the X, which is difficult to identify positively, is one of those in the second row. A female would have an XX chromosomal composition; the Y would be absent, and another X would replace it.** (Courtesy of Dr. Barbara Migeon.)

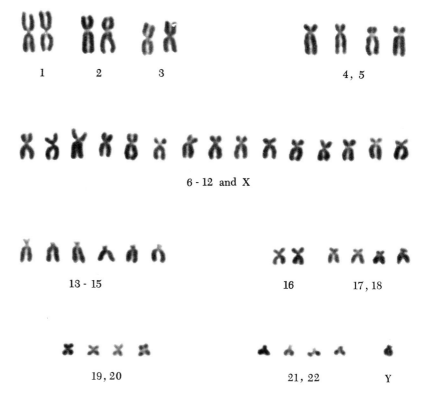

1   2   3   4, 5

6 - 12 and X

13 - 15   16   17, 18

19, 20   21, 22   Y

cannot go on indefinitely. At some time during the life cycle of an individual some compensatory mechanism must reduce this number, for we know that the cells of individuals belonging to the same species have a striking constancy of chromosome number. Thus normal human cells have 46 chromosomes; those of maize, 20; of the mouse, 40; of the rat, 42; and so forth. The lowest number known is 4, found in *Haplopappus,* a plant in the family Compositae, whereas some plants and animals have numbers as high as several hundred. This numerical constancy for each species is repeated generation after generation. The gametes, therefore, must have one-half the number of chromosomes found in the zygote and in the other cells of the body (since the latter arise from the zygote by mitosis). The reduction in number of chromosomes is accomplished by a special type of cell division called *meiosis,* which in its barest essentials consists of *two nuclear and cytoplasmic divisions but only one replication of chromosomes.*

Before considering the details of meiosis and the features that distinguish this type of cell division from mitosis, we need to recognize certain terms that conveniently describe chromosomal states. The chromosomes in the nuclei of gametes are variously said to be of *reduced, gametic, haploid,* or *n* number; those in the zygote and all cells derived from it by mitosis are termed of *unreduced, zygotic, diploid,* or *2n* number. Thus a human egg, prior to fertilization, possesses 23 chromosomes, in contrast to the 46 in the zygote. Furthermore, the 46 chromosomes are not all individually different; they exist as 23 pairs, as indicated in Figure 8.1, the members of each pair being similar in shape, size, and genetic content. The members of each pair are *homologous* to each other and *nonhomologous* with respect to the other chromosomes. In a zygote every pair of homologous chromosomes, or *homologues,* thus consists of one member contributed by the sperm and one by the egg.

One exception to the similarity of paired homologues in shape and size is the pair of chromosomes characterizing the two sexes. Figure 8.1 shows this pair of chromosomes in a human male. The human female is XX, and consequently her paired chromosomes are similar and homologous; the male is XY, and the two chromosomes are essentially nonhomologous.

Meiosis is a rather complicated type of cell division, yet the remarkable thing about it is that, like mitosis, the crucial nuclear events and end results are essentially the same wherever encountered. Consequently, a single account of it applies equally well to a fungus, an insect, a flowering plant, or a man. Except for the type of cell resulting from meiosis, the process is basically similar in both sexes as well.

We can separate meiosis into a sequence of steps similar to those in mitosis (Figures 8.2 and 8.3). Prophase, however, is a more leisurely process and hence longer in duration, and the modifications introduced affect the character of the resultant cells. Five separate prophase stages are recognizable.

Figure 8.2 **Diagrammatic representation of the states of division in meiosis I and II. For simplification, only one pair of homologues is included.** [M. M. Rhoades, *Journal of Heredity*, 41 (1950), 59–67.]

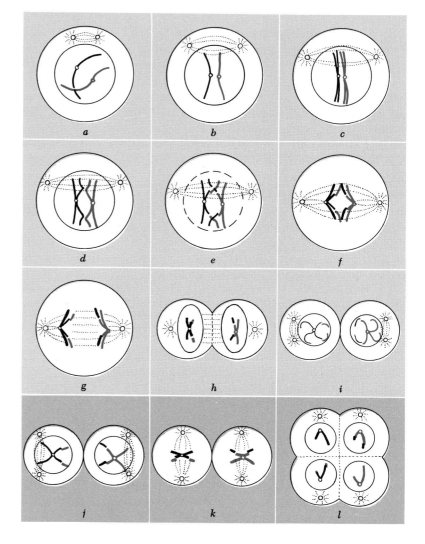

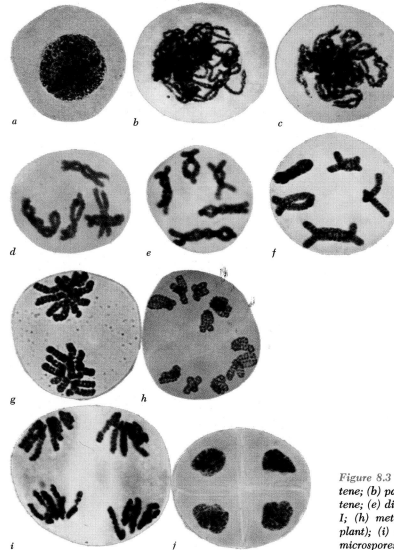

a   b   c

d   e   f

g   h

i   j

Figure 8.3   ***Stages of meiosis in*** Trillium: *(a) zygotene; (b) pachytene; (c) early diplotene; (d) late diplotene; (e) diakinesis; (f) metaphase I; (g) late anaphase I; (h) metaphase II (prophase II is absent in this plant); (i) anaphase II; (j) quartet stage, with four microspores.* (Courtesy of Dr. A. H. Sparrow.)

The *leptotene* stage initiates meiosis. Meiotic cells and their nuclei are generally larger than those of the surrounding tissues. The chromosomes, present in the diploid number, are thinner and longer than in mitosis, and are therefore difficult to distinguish individually. Leptotene chromosomes, however, differ from those in ordinary mitotic prophase in two ways: (1) they *appear* to be longitudinally single

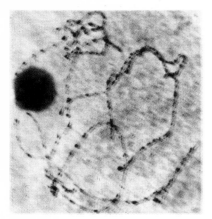

Figure 8.4 **Meiotic prophase (pachytene stage) in the wood rush,** Luzula, *showing the numerous chromomeres along the paired homologues, and the nucleolus with its attached chromosomes.* (Courtesy of Dr. S. Brown.)

Figure 8.5 **Zygotene stage of meiosis in the regal lily,** Lilium regale. *In the lower right-hand corner, both paired and unpaired regions of the homologues are visible.* (Courtesy of Dr. J. MacLeish.)

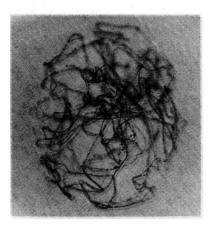

rather than double, although the timing of DNA synthesis indicates that they are in fact double; (2) their structure is more definite, with a series of dense granules, or *chromomeres,* occurring at irregular intervals along their length. The chromomeres of any given organism are characteristic in number, size, and position, and consequently can be used as landmarks to identify particular chromosomes, especially at pachynema where they are larger and more readily identified. Chromomeres are regions which have been compacted by localized coiling of the chromosome, and in this state the DNA contained within them is believed to be inactive metabolically, that is, it is not making RNA. It has been estimated that in the garden lily, for example, there are about 2,000 chromomeres in the entire set of 24 chromosomes, although in the plant *Luzula* the number is far fewer (Figure 8.4).

Movement of the chromosomes initiates the *zygotene* stage, and this movement results from an attracting force that brings together homologous chromosomes. The pairing of homologues, known as *synapsis,* begins at one or more points along the length of the chromosomes and then proceeds, much as a zipper would, to unit the homologues along their entire length. This is an exact, not a random, process, for the chromomeres in one homologue synapse exactly with their counterparts in the other (Figure 8.5). When synapsis is complete, the nucleus will appear as if only the haploid number of chromosomes is present. Each, however, is a pair of homologous chromosomes, and these are now referred to as *bivalents.*

Zygonema is the period of active synapsis. The next, or *pachytene,* stage is distinguishable by the fact that the paired chromosomes of each bivalent are easily seen (Figure 8.4), and since the chromosomes have continued to shorten and thicken by coiling, they are more readily identified one from the other. The chromomeres and the attachment of the nucleolus to a particular chromosome may be visible with high magnification (Figures 8.4 and 8.6).

The pachytene stage ends when the synaptic forces of attraction lapse and the homologous chromosomes separate from each other. This is the *diplotene* stage, and as Figure 8.7 indicates, each chromosome now consists of two chromatids. Each bivalent, therefore, is composed of four chromatids. Longitudinal replication of each chromosome took place prior to this stage, but did not become obviously evident until the attraction between homologues ceased.

Separation of the homologues, however, is not complete. At one or more points along their length, contact is retained by means of *chiasmata* (singular, chiasma). Each chiasma results from an exchange of chromatids between the two homologues; we shall discuss later in

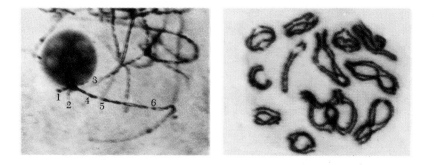

Figure 8.6 (left) **Pachytene in maize.** *Chromosome 6 of maize is shown here in its entirety, with the nucleolus attached at the nucleolar organizer region. The numbered regions are identifiable areas along the length of the bivalent; the paired region of the bivalent can be seen between regions 5 and 6. (Courtesy of Dr. B. McClintock.)*

Figure 8.7 (right) **Diplotene stage in a spermatocyte of a salamander,** Oedipina uniformis. *The two chromatids of each chromosome are readily visible, as are the chiasmata. The brushlike appearance of the bivalents is caused by thin loops of chromatin extending from the body of the bivalent. The heteromorphic bivalent (central upper left) consists of the X and Y chromosomes, with the Y being the shorter of the two. (Courtesy of Dr. J. Kezer.)*

this chapter the significance of this phenomenon as it relates to heredity, but the relationship of the chromatids to a chiasma is clearly indicated in Figure 8.8.

When only one chiasma has formed, the bivalent in the diplotene stage appears as a cross. If two are formed, the bivalent is generally ring-shaped; if three or more form, the homologues develop a series of loops. In different cells, the number and approximate positions of the chiasmata vary, even for the same bivalent, but, as a rule, long chromosomes have more chiasmata than short ones, although even the shortest seem to be able to form at least one chiasma.

The next prophase stage is called *diakinesis,* but the distinction between it and the diplotene stage is not a sharp one. During diaki-

Figure 8.8 **A single bivalent of** O. uniformis, *showing the crossing over of nonsister chromatids at the points of the two chiasmata. The centromeres of the two homologues appear as dark regions at the left of the bivalent. (Courtesy of Dr. J. Kezer.)*

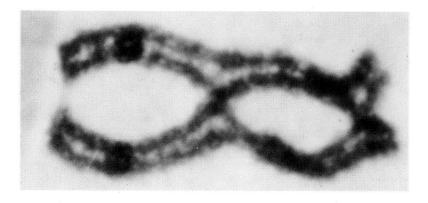

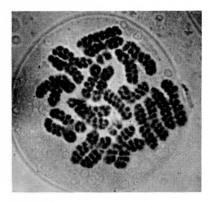

Figure 8.9 **Coils as they appear in the meiotic chromosomes of the spiderwort,** Tradescantia paludosa, *at anaphase. The chromosomes are flattened onto a single plane by the squash technique.*

nesis the nucleolus becomes detached from its special bivalent and disappears, and the bivalents become considerably more contracted. Also as contraction proceeds (Figure 4.6), the chiasmata tend to lose their original position and move toward the ends of the chromosomes.

We have mentioned that the chromosomes shorten as they progress from the leptotene stage onward through prophase. This is accomplished by the development of a series of coils that gradually decrease in number as their diameters increase. The process is no different from the shortening of chromosomes in mitosis; the coils here, however, are more easily observed, particularly when the cells have been pretreated with ammonia vapors or dilute cyanide solution before staining. Figure 8.9 illustrates the coils as they appear in the spiderwort, *Tradescantia.*

The breakdown of the nuclear membrane and the appearance of the spindle terminate prophase and initiate the *first metaphase of meiosis* (Figures 8.2 and 8.3). The bivalents then orient themselves on the spindle, but instead of all centromeres being on the equatorial plate, as in mitosis, each bivalent is so located that its centromeres lie on either side of, and equidistant from, the plate (Figure 8.10). This seems to be a position of equilibrium.

The *first anaphase* of meiosis begins with the movement of the chromosomes to the poles (Figures 8.10 and 8.11). The two centromeres of each bivalent remain undivided, and their movement to the opposite poles of the spindle causes the remaining chiasmata to slip off and free the homologues from each other. When movement ceases,

Figure 8.10 **Meiosis in the male milkweed bug,** Oncopeltus fasciatus. *(a) Metaphase I, with the bivalents oriented on the metaphase plate (the spindle is not stained and hence is invisible); (b) metaphase II, during which the chromatids will segregate; (c) anaphase I. (Courtesy of Dr. S. Wolfe.)*

a   b   c

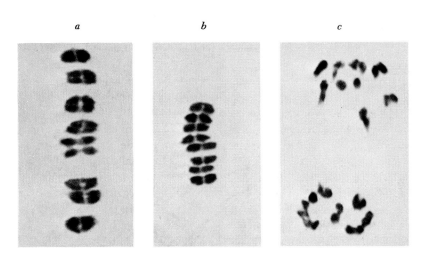

a reduced, or haploid, number of chromosomes will be located at each pole. Unlike mitotic anaphase, in which the chromosomes appear longitudinally single, each chromosome now consists of two distinctly separated chromatids united only at their centromeres. The nucleus then forms, the chromosomes uncoil, and the meiotic cell is bisected by a membrane wall. This is the *first telophase of meiosis* (Figure 8.2).

After an interphase which, depending on the species involved, may be short or long—it may even be absent altogether—the chromosomes in each of the two haploid cells enter the *second meiotic division* (Figures 8.2 and 8.3). If an interphase is absent, the chromosomes pass directly from the first telophase to the *second prophase* without any change in appearance (Figure 8.3). If an interphase is present, a nuclear membrane forms in telophase, the chromosomes uncoil, and a somewhat more prolonged second prophase is found. But whatever the case, the chromosomes reaching the *second metaphase* (Figure 8.10) are essentially unchanged from what they were in the previous anaphase; that is, *no chromosomal replication occurs during interphase,* and the centromere of each chromosome remains functionally undivided. A spindle forms in each of the two cells, and, at the *second anaphase,* the centromeres separate and the chromosomes move to the poles. The nuclei are reorganized during the *second telophase,* giving four haploid nuclei that become segregated into individual cells by segmentation of the cytoplasm.

Looking back over the events of meiosis, we find that the chromosomes remained unchanged in longitudinal structure from the diplotene stage to the end of the second meiotic division. The replication of each chromosome occurred during the premeiotic interphase, but this was followed by two divisions; in the first the homologues separated from each other to reduce chromosome number, an event made possible because synapsis joined them; in the second the two chromatids of each chromosome separated.

At this point you may well ask why the reduction in chromosome number could not be accomplished just as efficiently with a single division instead of two. Where only a single meiotic division is found, as happens during sperm formation in the normally haploid male honeybee, and where a reduction in chromosome number is not a necessary feature in the life cycle, it is essentially like the second rather than the first division, and therefore more mitotic in character except for the nature of the resultant cells. In organisms that have a diploid chromosome number, the reduction could take place in one division if no prior replication of the chromosomes had occurred, but since DNA replication appears to be part of the initiating mechanism

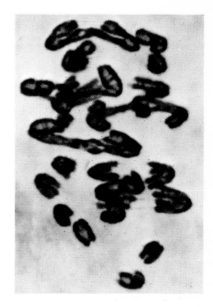

Figure 8.11   Anaphase I of spermatogenesis in the salamander O. uniformis. *(Courtesy of Dr. J. Kezer.)*

of division, the second meiotic division, without another round of replication, is necessary to bring about the reduction in chromosome number.

THE PRODUCTS OF MEIOSIS

In the animal kingdom, meiosis leads to the formation of sexual gametes, the egg and sperm usually being the only cells carrying a haploid complement of chromosomes. In the plant kingdom, however, meiosis can occur at various times during the life cycle, and the haploid products may be sexual gametes or asexual spores, depending on the particular group of plants being studied. Since the plant life cycles are covered in another volume in this series,* we shall consider here only the products of vertebrate meiosis, that is, the egg and sperm.

The primordial germ cells of the human embryo make their appearance approximately 20 days after fertilization, and migrate from their origin in the wall of the yolk sac to the developing gonads during the fifth week of development. Once located in the female gonad, these cells become the source of the female germ cells. They divide rapidly to form clusters of *oogonia* near the outer wall of the ovary, and each cluster is transformed into a layer of flat epithelial cells surrounding a central cell which, at the end of the third month of development, becomes the *primary oocyte*. The cluster is known as the *primordial follicle*. The primary oocyte enters meiosis as soon as it is formed, and by the seventh month of development, all of the oogonia have disappeared, and the oocytes have reached the so-called *dictyotene* stage which follows pachynema and in which the chromatin is quite diffuse in appearance. The oocytes remain in this state until sexual maturity.

By birth, therefore, a human female will have formed all of the oocytes she will have and they will have progressed well into meiosis. The number of oocytes has been estimated to range in number from 40,000 to 300,000. During each ovarian cycle, several oocytes begin development, but only one achieves maturity; the remainder disintegrate. As a rule, therefore, only a single functional oocyte produces a fertilizable egg during each ovarian cycle, and if the childbearing years are assumed to cover the age period of 12 to 50, only 400 or so of the oocytes reach maturity. The entire course of events from primordial germ cell to *Graafian follicle* is depicted in Figure 8.12.

* H. C. Bold, *The Plant Kingdom* (2nd ed.) (Englewood Cliffs, N.J.: Prentice-Hall, Inc., 1964).

Initially, the *primary oocytes* lie close to the germinal epithelium, but later they increase in size and sink into the interior of the ovary where they become surrounded by *follicle cells,* which probably have both a protective and a nutritive function. The whole structure is now known as a *Graafian follicle.* During this process of enlargement and encapsulation, the oocyte is building up reserve food material, the yolk. This food, which may be protein or fat, in mammals is generally distributed throughout the cytoplasm as yolk spheres or granules. In the frog, however, the yolk so completely fills the cell that the cytoplasm is restricted to a small fraction of the cell surrounding the nucleus; the well-known yolk in the hen's egg is also enormous compared to the amount of cytoplasm.

Eventually, the Graafian follicle ruptures and the egg (Figure 6.3), or *ovum,* is released from the ovary, and passes into the *oviduct,* or *Fallopian tube,* where it can be fertilized by a sperm. By this time, however, meiosis has been resumed and has reached metaphase of the second meiotic division. Meiosis will be completed only if the egg is fertilized, the sperm acting as initiating agent. Only a *single* functional cell results, however. The other three cells, or *polar bodies,* are cast off and will degenerate, *but the process has effectively reduced the chromosome number without depriving the egg of the cytoplasm and yolk the embryo will need when it begins to develop.*

The first meiotic division in the primary oocyte takes place close to the cell membrane, and the outermost nucleus, together with a small amount of cytoplasm, is pinched off as a polar body (Figure 8.13). The second meiotic division results in the pinching-off of a

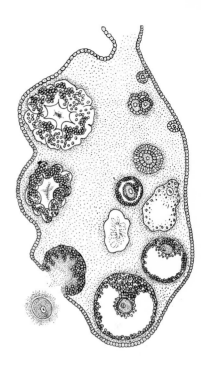

Figure 8.12 **Section of mammalian ovary,** *showing the progressive development of the oocytes as they arise from the germinal epithelium, increase in size, sink into the interior of the ovary wall, and finally escape to the outside by rupture of the wall of the Graafian follicle.*

a *b*

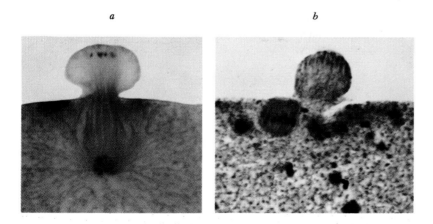

Figure 8.13 **Polar-body formation in the egg of the whitefish,** *Coregonus. (a) Anaphase of the first meiotic division, with the first polar body being pinched off. (b) Metaphase of the second meiotic division, which will lead to the pinching off of a second polar body. In the meantime, the first polar body may also divide to give a total of three polar bodies.* (General Biological Supply House, Inc.)

second polar body; the first polar body, meanwhile, has also undergone a second meiotic division, thus giving a total of three polar bodies. The haploid nucleus remaining in the egg is now known as the *female pronucleus.* It sinks into the center of the cytoplasm and is ready for union with a similar haploid nucleus brought in by the sperm during fertilization.

The primitive germ cells of the human male enter the developing gonad during the fifth week of development. They become incorporated into the *sex cords,* which at first are solid structures but which after birth develop a lumen and become the *seminiferous tubules.* These make up about 90 percent of the bulk of the testes.

The germinal epithelium contains *spermatogonia,* cells that continue to increase their number by mitotic division until senility sets in. These derivative cells mature into *primary spermatocytes,* which undergo a first meiotic division to produce *secondary spermatocytes;* the latter pass through a second meiotic division, giving four cells called *spermatids.* These become motile sperm by a remarkable transformation of the entire cell. The human male differs from the female in that spermatogonia persist and continue to produce primary spermatocytes, and hence viable sperm, from the beginning of sexual maturity to old age.

The mature sperm consists essentially of a head and a tail. The head is a highly compacted nucleus, capped by a structure known

Figure 8.14 **Transformation of the Golgi complex (a) into the acrosome (b) during the process of spermiogenesis in the house cricket,** Acheta domestica. *The paired membranes of the Golgi complex are regularly spaced although several vacuoles are enclosed by the membranes. The sperm head at the right shows the dark, solid nucleus capped by the acrosome; the cone-shaped acrosome is formed by the Golgi complex, but the membranes of the Golgi complex are eventually sloughed off and do not become part of the acrosome.* (Courtesy of Dr. J. Kaye.)

a   *b*

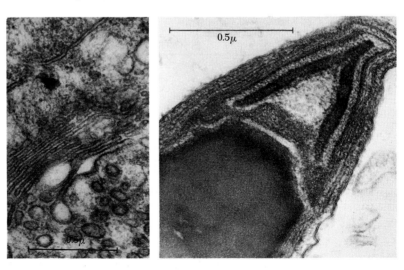

as the *acrosome* (Figures 6.3 and 8.14). It is derived from the Golgi materials of the spermatid, and apparently functions as a device for penetrating the egg during fertilization. Just back of the compacted nucleus is the *middle piece*, formed by an aggregation of the mitochondria. It develops as a sheath around the filament, or tail, and provides the tail with energy for locomotion. The filament, in turn, has developed as the result of a tremendous growth from one of the centrioles; the other centriole remains just beneath the nucleus, and at the time of fertilization enters the egg along with the male nucleus. Virtually no cytoplasm except particulate structures is used to form the mature sperm.

Each spermatid, therefore, has been transformed from a rather undifferentiated cell into a highly specialized cell capable of reaching the egg under its own power, and of penetrating it once it has made contact.

The mature egg and sperm must unite with each other within a limited period of time, for neither has an indefinite life span. The critical period may be a few minutes, or it may be spread over several hours or days. In mammals, fertilization can occur periodically as the egg leaves the ovary and passes down the *oviduct* on its way to the *uterus*. Insects, however, mate only once, and sperm are stored in the female and used throughout the entire egg-laying period; in the honeybee, for example, this period may last a year or more. It is now possible to store mammalian sperm for an indefinite period by freezing them, and by means of *artificial insemination* the sperm of a single sire may be used to fertilize the eggs of many females. This practice has been widely used in animal breeding programs, thus passing on the superior qualities of one sire to many offspring, and it has been successfully carried out in humans when, for one reason or another, normal conception fails.

The essential process of fertilization is the union of male and female pronuclei, but the sperm also functions as an activating agent. That is, nature has insured against the egg beginning its embryonic development in an unfertilized state; if it did, haploid embryos would result and even if such embryos were viable and developed into sexually mature adults, the process of meiosis would be hopelessly complicated. Unfertilized eggs of mammals and other related vertebrates can be induced to initiate development by various artificial means, but this rarely occurs naturally.

Fertilization is also a specific process in that the sperm of one species will not, as a rule, fertilize the egg of another species. It now appears that several chemicals are present to insure proper fertilization and to prevent the penetration of foreign sperm. The egg produces a protein substance called *fertilizin* which reacts with an *antifertilizin* on the surface of the sperm; fertilizin may act to attract sperm of its own kind, but once the two substances interact, the sperm becomes firmly attached to the egg membrane, and is then drawn into the interior of the egg. Other sperm are barred from entry by the changes that then take place in the *vitelline membrane* of the egg, an outer coating found on most eggs.

Only the nucleus and one centriole of the sperm enter the egg. The former fuses with the female pronucleus, the latter divides and begins formation of the first division spindle. In summary, therefore, the entry of the sperm into an egg contributes: (1) a stimulus to development; (2) a set of haploid chromosomes, which is the paternal hereditary contribution to the newly formed zygote; and (3) a centriole, which is involved in the machinery of cell division.

GENETIC SIGNIFICANCE OF MEIOSIS

We have presented meiosis as a logical and necessary part of the life cycle of a sexually reproducing organism, that is, it is the opposite of fertilization as regards the number of chromosomes. So far as heredity is concerned, we need to clarify two additional implications.

Figure 8.15 illustrates the segregation of chromosomes, with the paternal chromosomes indicated in black, the maternal ones in color. Each bivalent at the first meiotic metaphase would, of course, consist of two homologues, one from each parent. The orientation of all bivalents on the spindle is usually random, so segregation at anaphase leads to a random distribution of chromosomes. The haploid cells resulting would therefore contain a mixture of paternal and maternal chromosomes. When 4 pairs of chromosomes are involved, 16 different combinations of 4 haploid chromosomes are possible. The number possible can be readily determined by calculating the value of 2^n, when n equals the number of pairs of chromosomes. In man, who has 23 pairs, the number of possible gametic chromosome combinations is 2^{23}, or 8,388,608. The chance of any single human sperm or egg containing only paternal or maternal chromosomes is, therefore, negligible.

The distribution of paternal and maternal chromatin to their offspring through gametes is further complicated by the process of chiasma formation (Figures 8.7 and 8.8). As we pointed out before, a chiasma results from an exchange between chromatids in the two homologues. One of these is from a maternal chromosome, the other from a paternal chromosome. If we further consider that the chromosome consists of a number of genes strung along its length, and that the genes in one homologue may be slightly different from those in the other as the result of mutations, a situation such as that illustrated in Figure 8.16 can be envisaged.

The genes in the paternal (black) chromosome are designated by capital letters from *A* to *E*; those in the maternal (color) one from *a* to *e*. Remember, however, that any single chromosome may have hundreds of different genes along its length. Chiasmata have formed as the result of exchanges between *A* and *B*, and between *C* and *D*. (In genetic terminology, an exchange of chromatids is called *crossing over*, and the genes *A* to *E* and *a* to *e* would constitute *linkage groups*.) The actual mechanism responsible for chiasma formation is not known for certain, but it seems clear that it occurs before the diplotene stage when the chromosomes have a clearly demonstrable double longitudinal structure. The important thing, however, is that chiasma formation breaks up linkage groups, and therefore alters the set of genes the chromosome possessed before entering meiosis. Since the chromatids eventually are distributed to the four haploid cells, it is clear that each gamete is genetically different from the others.

We see, therefore, that both the random segregation of paternal and maternal chromosomes and the breaking up of linkage groups through chiasma formation insure that the haploid cells resulting from

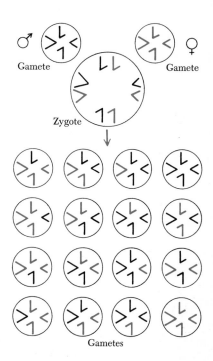

Figure 8.15 *The random segregation of paternal (black) and maternal (color) chromosomes during meiosis. Crossing over and linkage are not indicated.*

Figure 8.16 *The genetic consequences of crossing over. (a) A bivalent, consisting of paternal (black) and a maternal (color) homologue, has formed, and crossing over has taken place between genes A and B, and C and D. (b) At anaphase the two chromatids in each segregating chromosome are no longer alike genetically. (c) The chromatids are now separated, and two of them have a different genetic composition, while the other two remain as before.*

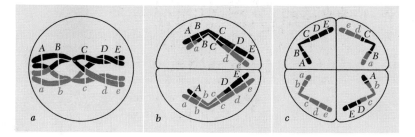

meiosis will have a variable combination of genes. Since these cells contribute through fertilization to the next generation, the individuals of that generation must exhibit a comparable genetic variation. It is this inherited variability which natural selection acts on to bring about the evolution of organisms. Sexual reproduction, with its complementary phenomena of fertilization and meiosis, is a means not only for the production of new individuals, but of new individuals *that vary among themselves*. In this sense meiosis differs greatly from mitosis, which, in its production of similarly endowed cells, is a conservative process of reproduction.

IN THE PAST FEW CHAPTERS, WE HAVE CASUALLY MENTIONED THAT organisms "develop" from a fertilized egg into a plant or animal of adult proportions. Each of us knows in a general way what is meant by development: it is a continuous and gradual process of change that takes time in order to be fully realized; it is generally accompanied by an increase in size and weight; it involves the appearance of new features and new functions; and it eventually slows down when mature dimensions are reached. Man, for example, develops from the fertilized egg stage through embryonic and prenatal life, childhood, adolescence, sexual maturity, physical maturity, middle age, senility, and death. Development is, of course, one of the most prominent features in the early life of an organism, but the formation of new blood cells, gametes, and wound tissue, which may take place up to death at an advanced age, are also aspects of development. So, too, are those processes we associate with aging, for example, excess formation of collagen in the extracellular spaces and the calcification of joints. These, it would appear, are normal processes of development continuing beyond the point of a functional and developmental optimum. The terms we have used, however, are only broadly descriptive. They tell us very little about the mechanism of development as a biological phenomenon. For that, we approach development from the cellular level, since the cell remains the building block of life. We must ask, for example, how the potentialities of the fertilized egg, which reside as coded information in the egg's DNA and in the organization of its cytoplasm, can become the fully realized features of a perfectly formed organism, with each organ being the right size, in its allotted place, and equipped with the cells needed to perform properly.

The problem of development, in its entirety, is a complicated one, and many questions remain unsolved. Development is also an extraordinarily precise process. Consider the striking similarity of identical twins. Each twin consists of thousands of billions of cells, and in the course of time the processes of growth and differentiation form and mold these cells into a unique individual. Yet so precise are the patterns of developmental change that even minor as well as major physical, physiological, and behavioral characteristics emerge and are expressed by both twins. Since identical twins have their origin from a single fertilized egg, the control mechanisms must have been present in the fertilized egg, and must initially have been operative at a cellular level before the pattern becomes expressed at higher levels of organization. From what has been stated previously, this means that different kinds of cells achieve their distinctness because they possess distinct sets of proteins, and distinct sets of proteins, in turn, are determined by the particular set of genes operating in each cell.

Growth is defined as an increase in mass. This increase can result solely from an enlargement of cells, but more often it is accompanied by an increase in the number of cells through mitotic divisions. Growth, then, is essentially a process of replication: the original cell takes from its environment the raw materials it needs and converts them into more substance and more cells like itself. Let us consider the human egg. It weighs about 1×10^{-6} g, and the sperm, at fertilization, adds to it only another 5×10^{-9} g. At birth, however, a child will weigh around 7 lb, or 3,200 g, which is an increase of about one billion times during the 9-month prenatal period. A newborn child is obviously not simply a mass of cells of comparable size and character to the original cell; if it were, it would be just a ball of cells devoid of human qualities. Nor has its growth rate been uni-

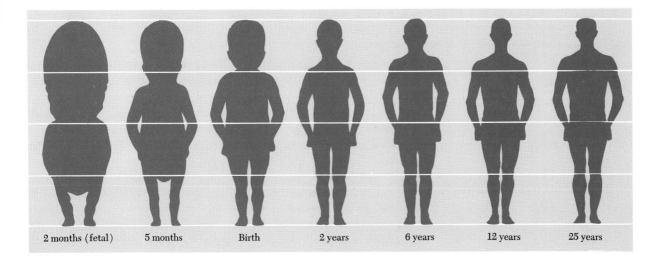

Figure 9.1   Changes in the form and proportion of the human body during fetal and postnatal life. [From H. B. Glass, *Genes and the Man* (New York: Columbia University Teachers College Bureau of Publications, 1943), after Morris.]

| 2 months (fetal) | 5 months | Birth | 2 years | 6 years | 12 years | 25 years |

form throughout its prenatal life. Other processes must act to mold cells into shape, as a potter or a sculptor molds his clay, and to stamp them with character.

One of these processes is the *relative rate of growth*. This rate helps to determine form, which is another way of saying that some parts of the body grow at a faster or slower rate than others, and that in development some features come into existence early, and others late. Figure 9.1 shows how the growth rate in the human being alters the relative proportions of bodily parts to one another. The head and neck increase in size rapidly during the early period of gestation, the arms grow faster at an earlier stage than do the legs, whereas the trunk progresses at a more or less steady rate until maturity.

Growth, therefore, is not just the enlargement and multiplication of cells; it is a complicated pattern, with different centers of growth being active at different times and at different rates of development. These centers are coordinated to produce an unfolding of form, and it is form as well as function, of course, that distinguishes man from other animals, one human being from another, and an orchid from a lily. We shall return to a consideration of form later in the chapter.

THE CELL IN DEVELOPMENT

It is *differentiation*, however, that stamps each cell with its own uniqueness of structure and function. A generalized cell is gradually transformed by a process of progressive changes into a specialized one, and variation is thereby introduced into a functioning organism. In a unicellular organism, differentiation alters the character of only a single cell; among multicellular organisms, the change is both within and among cells. In man, for example, growing cells are transformed into the myriad of different cells that makes up the human body (Figure 9.2): cells of the nervous, muscular, digestive, excretory, circulatory, and respiratory systems.

Differentiation, therefore, is a process of directed change. It is a phenomenon that has no counterpart in the nonliving world, and what information about it we have has been derived from observations of living systems. This process is creative in the sense that life is creative, for out of the general features common to all cells arise structures and functions that are peculiar to specialized cells. Specialization can

Figure 9.2 **Differentiation of generalized mesoderm cells (mesenchyme) into two kinds of specialized cells, muscle and cartilage.** Both these cells are similar in that they produce substantial amounts of protein, but in muscle cells the proteins (actin and myosin) are retained internally for contractile purposes, while in cartilage cells the protein is deposited as collagen outside the cell, where it plays a supportive role. The ultimate shape of the cells also changes as a result of differentiation. [After C. H. Waddington, *Principles of Development and Differentiation* (New York: The Macmillan Company, 1966).]

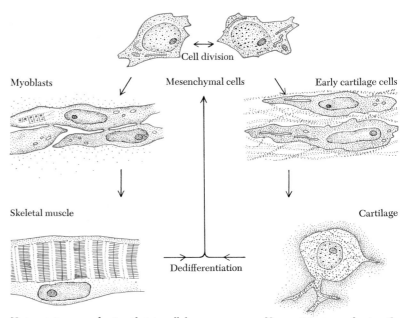

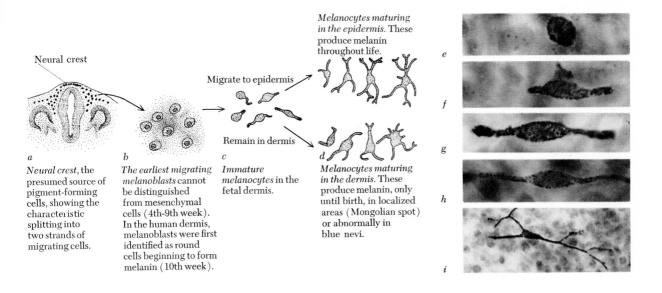

Melanocytes maturing in the epidermis. These produce melanin throughout life.

Neural crest

Migrate to epidermis

Remain in dermis

a — *Neural crest*, the presumed source of pigment-forming cells, showing the characteristic splitting into two strands of migrating cells.

b — *The earliest migrating melanoblasts* cannot be distinguished from mesenchymal cells (4th-9th week). In the human dermis, melanoblasts were first identified as round cells beginning to form melanin (10th week).

c — *Immature melanocytes* in the fetal dermis.

d — *Melanocytes maturing in the dermis.* These produce melanin, only until birth, in localized areas (Mongolian spot) or abnormally in blue nevi.

e
f
g
h
i

Figure 9.3   ***The origin and course of differentiation of human melanin-producing cells.** (a–d) The melanoblasts have their origin during the fetal stage in an area called the neural crest, from which they migrate and differentiate. Their ultimate fate depends on whether they end up in the dermis or epidermis; if in the former, they cease producing melanin at birth except in special circumstances, but if in the epidermis, they continue to produce melanin until death. (e–i) Five cells, showing the series of changes that are undergone from the round melanoblast (e) to the highly differentiated melanocyte (i).* [Courtesy of Dr. A. A. Zimmermann, from A. A. Zimmermann and S. W. Becker, Jr., *Illinois Monographs in Medical Sciences*, VI (1939), 1–39.]

be seen in Figure 9.3, which illustrates the origin and progressive differentiation of a special cell type, in this case, the *melanocytes* that form pigment in the human skin. Differentiation, therefore, is to development what mutation is to biological inheritance, and what imagination is to human endeavor; it provides variety of form, function and behavior, but without at any time destroying the unity of an organism as an individual.

Let us examine the process of melanocyte formation since, at a readily visible level, it represents an excellent example of differentiation within the cell. The melanocytes have their orgin as *melanoblasts* in a region of the embryo called the *neural crest,* from which they migrate to the outer (*epidermis*) or inner (*dermis*) layers of the skin. As they migrate, they alter their shape, as Figure 9.3 indicates, and they also begin to form small granules, *melanosomes*, within which the pigment *melanin* is bound.

Figure 9.4 illustrates the steps in the differentiation of melanosomes. Originating from the free ribosomes in the cytoplasm are slender fibrils, presumably protein, that aggregate into larger and larger fibers. Melanin accumulates on these fibers, eventually obscuring the fine structure, and the fibers are then gradually enclosed within a membranelike envelope. The mature melanosome is an electron-dense body showing no internal structure.

Many known genes influence the coloring of vertebrate animals. A study of these genes in correlation with the developing melanosome shows that they influence the number and arrangement of the fibers, the character and distribution of melanin in the melanosome, and the size, shape, and distribution of melanosomes. The system, therefore, is a beautiful example of how genes in the nucleus bring about changes within the organelles of the cytoplasm during the course of differentiation.

If we contrast differentiation with growth, we find, without knowing why, that these processes tend to be mutually exclusive, although various stem cells and liver parenchyma divide even though differentiated. Where growth is, to varying degrees, an unending process of multiplication of similar units, differentiation is the extraction and modification of a unit from the mass, thus making it distinctive. In the process, differentiation tends to prevent the further multiplication

Figure 9.4 **Pattern of differentiation of a melanosome.** *(a–d) Four stages of development, with fiber formation within the matrix sheet (a) and then the progressive accumulation of melanin on the fibers until the melanosome appears as a solid structure (b–d). (e) Detailed interpretation of the origin of the protein fibers from the ribosomes of the cytoplasm (small solid black bodies) and the grouping of six fibers into a compound fiber.* (Courtesy of Dr. F. Moyer.)

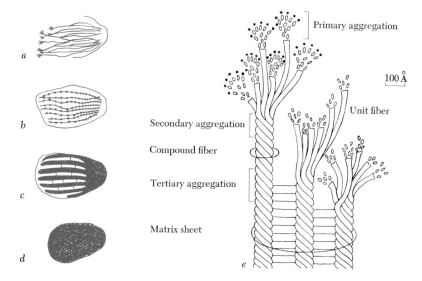

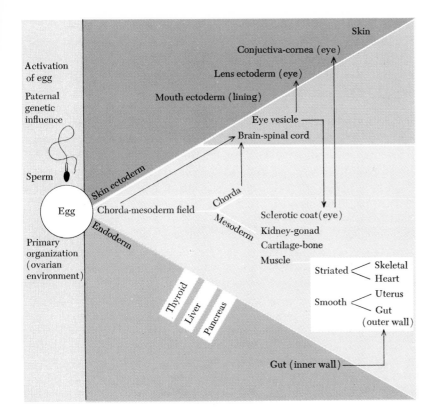

Activation
of egg

Paternal
genetic
influence

Sperm

Egg

Primary
organization
(ovarian
environment)

Skin ectoderm

Chorda-mesoderm field

Endoderm

Skin

Conjuctiva-cornea (eye)

Lens ectoderm (eye)

Mouth ectoderm (lining)

Eye vesicle

Brain-spinal cord

Chorda

Mesoderm

Sclerotic coat (eye)

Kidney-gonad

Cartilage-bone

Muscle

Striated — Skeletal
          — Heart

Smooth — Uterus
         — Gut
           (outer wall)

Thyroid
Liver
Pancreas

Gut (inner wall)

Figure 9.5 **Diagrammatic representation of the pattern of progressive
differentiation from unfertilized egg
to mature tissues in a vertebrate. The
three major tissue layers (ectoderm,
mesoderm, and endoderm) originate
early and progressively give rise to
the cells of the major organs. Dashed
lines indicate an influence of one tissue on another during the course of
development. Note that the eye has
a double origin from both ectoderm
and mesoderm. (Courtesy of Dr. B. H.
Willier.)**

of the cell. Therefore, the more differentiated a cell has become, the
less likely it is to divide. The cell thus becomes committed to a course
of action it cannot readily change.

Let us examine a bit more closely what we mean when we say
a cell is "committed." Figure 9.5 shows diagrammatically the course
of development in vertebrate animals. When the cells near their final
form as mature, differentiated structures, their potentiality for further
change narrows as their specialization becomes more pronounced.
C. H. Waddington, an English embryologist, has expressed this idea
of commitment by his diagram of a developmental landscape (Figure 9.6). He visualizes a generalized cell as a ball rolling downhill
toward its final destiny, a destiny that depends on which of the
many valleys the ball rolls through. The further the cell penetrates
into the developmental landscape, the greater is the loss of general
properties and the greater the acquisition of special features, and the
less likely is it able to return to an undifferentiated state.

THE CELL IN DEVELOPMENT

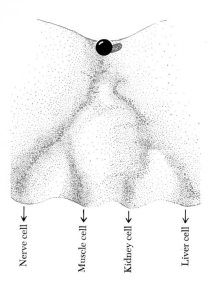

Figure 9.6 *How an uncommitted cell (represented by a ball) may become committed by rolling down one of the channels of differentiation.* [After C. H. Waddington, *The Strategy of the Genes* (New York: The Macmillan Company, 1957).]

Nerve cell ← Muscle cell ← Kidney cell ← Liver cell ←

Let us express what we have been saying in more specific terms. Figure 9.7 illustrates the general course of early development in an animal such as amphioxus. If we shake apart the four-celled stage, we find that each cell is capable of developing into a normal, if slightly smaller, organism; such development will occur even if we pinch the *blastula* in two with a hair loop. However, if we wait until the *gastrula* stage, the potentiality of cells within the embryo has begun to vary as a result of differentiation. The initial change is nuclear, and by nuclear transplantation it is possible to show that some nuclei, placed in an enucleated egg, can give rise to normal individuals while only abnormal and incompletely formed individuals will result from other such transplanted nuclei (Figure 9.8). Something obviously has been lost by certain cells of the gastrula that was present in earlier ones. Thus the capability of some cells has already been determined by the gastrula stage even though no obvious change has taken place, and from this point on, each cell can perform only its "committed" role.

An embryologist has another way of approaching this problem. He cuts certain cells out of an embryo and transplants them to other embryos (Figure 9.9). If he transplants a group of young, undifferentiated cells to the future head region of another embryo, the transplanted cells become part of the head region; if he transplants them to the back, they will become part of the back musculature; if to the posterior part, they become part of the tail. But if he transplants "committed" cells from an older embryo in the same way, instead of becoming an integral part of the region to which they are transplanted, they tend rather to retain their own identity and even to modify the

Figure 9.7 *Early developmental stages of an amphioxus from the egg (a) through the blastula (h) to the gastrula (j) stages. Although the size of the embryo remains much the same until gastrulation (infolding), cell size and shape are being altered constantly by division and the pressure of adjoining cells.* [Reprinted with permission from R. Gerard, *Unresting Cells* (New York: Harper & Row, Publishers, 1949).]

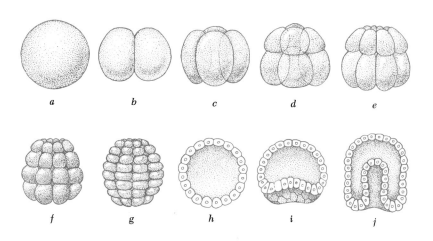

a   b   c   d   e

f   g   h   i   j

THE CELL

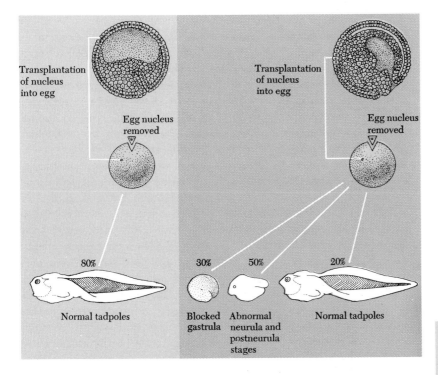

Figure 9.8 *A comparison of the ability of two kinds of nuclei to induce development when transplanted into enucleated eggs. Only about 40 percent of the eggs develop regardless of the source of the nuclei, but in those derived from peripheral cells, 80 percent of the developing eggs form normal tadpoles (left), while only 20 percent are normal when the nuclei are obtained from cells in the interior of the gastrula. The nuclei are, therefore, varied in their potentiality as a result of differentiation.* [After L. J. Barth, *Development: Selected Topics* (Reading, Mass.: Addison-Wesley Publishing Co., Inc., 1964).]

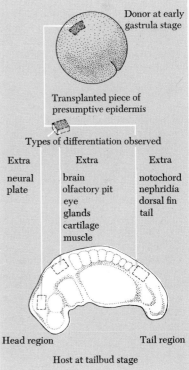

Figure 9.9 *An experimental demonstration that cells that would ordinarily form epidermis can, if transplanted to other regions of an older embryo, be incorporated into those regions. Differentiation, therefore, occurs after the transplantation and not before.* [After L. J. Barth, *Development: Selected Topics* (Reading, Mass.: Addison-Wesley Publishing Co., Inc., 1964).]

129

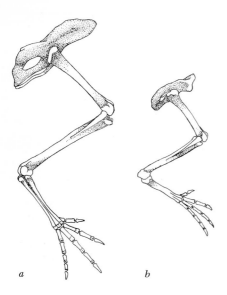

a   *b*

Figure 9.10 An example of a structure that develops after transplantation in a reasonably normal fashion. (a) Normal leg bones of a chick 18 days after incubation. (b) A slightly smaller but reasonably complete set of leg bones that developed after the hind limb bud (similar to the limb buds shown in Figure 10.4) was transplanted to the body cavity. At the time of transplantation, the limb bud showed no evidence of bone or muscle, but the cells had already been "committed" to leg formation, a process of differentiation that continued even though the limb bud had been removed to a foreign location. [From V. Hamburger and M. Waugh, Physiological Zoology, XIII (1940), 367–380.]

surrounding cells. This is well illustrated by an experiment done in the chick embryo. If the leg bud, which has no resemblance to a mature leg in any way, is removed from a young chick embryo and is transplanted to the body cavity of another embryo, the cells in the bud live, continue to increase in number, and eventually form therein a very well-developed leg with bones and muscles (Figure 9.10). Yet the bud at the time of transplantation had no obvious bone or muscle cells. In terms of Waddington's landscape, however, the cells had already entered a "valley" leading to leg formation, a valley down which they continued to roll and from which they could not escape.

We can see, therefore, that differentiation occurs during the course of time. But there is also a spatial aspect of differentiation. This can be determined by experiments on the limb buds of amphibians (Figure 9.11). There are two main axes of polarity to a limb: an *antero-posterior* axis that relates the limb to the head and tail of the animal, and a *dorsoventral* axis that relates to the top and bottom of the animal. If an early limb bud is excised and implanted, but rotated in such a way as to reverse its position, the limb will form. Its antero-posterior relation, however, will be reversed. If a limb bud from a later stage is handled in the same manner, not only will the antero-posterior relationship be reversed, but so too will be the dorsoventral relationship. In terms of spatial differentiation, therefore, the antero-posterior polarity is fixed early, and the dorsoventral polarity at a later time of development.

Differentiation, consequently, begins long before any visible morphological change takes place in the cells. Such changes, of course, must be preceded by chemical alterations that the cell apparently cannot easily undo once they have occurred. We must admit, however, that our present knowledge of these changes and the cause of their initiation is fragmentary indeed. There seems to be no doubt, on the other hand, that the differentiation of cells can occur either through the *loss of old functions* or through the *acquisition of new ones*. We should now examine these possibilities in somewhat greater detail, for loss and acquisition may not be quite so opposite as the words imply.

A mature nerve cell traces its ancestry back to the original fertilized egg, although it is, of course, very different from the egg both in morphology and in function. Irreversible changes occur along the way, and the early versatility of the egg is sacrificed for the special property of conduction that characterizes nerves. Since if a nerve is cut the axon can be regenerated, it has evidently not lost all power of repair. On the other hand, it cannot divide to form new nerve cells. If a nerve cell dies, therefore, it cannot be replaced, but this

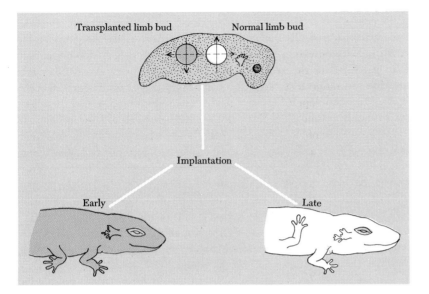

Figure 9.11 *Tissue polarity in limb buds. The foreleg is the normal limb; the hind limb developed from an implanted limb bud. When implanted early, the bud, when reversed at the time of implantation, shows a reversal only in the anteroposterior axis; that is, the "thumb" digit points rearward, and the "elbow" forward. When implanted at a later stage, both the anteroposterior axis and the dorsoventral axis are reversed; that is, the thumb points to the rear, and the limb as a whole is directed upward in growth instead of down. The two axes, therefore, are determined at different times during development.* [After C. H. Waddington, *Principles of Development and Differentiation* (New York: The Macmillan Company, 1966).]

loss in the power of division is compensated for by the acquisition of a new ability, namely, the capacity to conduct electrical impulses rapidly and efficiently.

We could, of course, assume that the original egg was capable of doing all the things that each differentiated cell can do, and that these abilities, residing in various parts of the cell, were segregated out by cell division. This possibility seems unlikely, however, when we consider that each of the cells of an early embryo can, when shaken apart, form normal individuals. A more likely supposition is that the egg contains, in the coded information in its DNA, the potentiality of all properties of all cells, and that as the differentiation of cells takes place, some potentialities are initiated by the activation of dormant genes while others are suppressed by gene inactivation. A consideration of the development of a chicken heart will make this clear.

The chicken heart begins to appear as a morphologically visible structure about 24 hours after incubation, and the first heart beats occur early in the second day. The cells that form the heart migrate from other regions into the heart-forming area, as indicated in Figure 9.12, but what we want to know is when the heart-forming cells first become identified as potential heart cells. To find this out we determine at what stage the cells begin to form chemical substances peculiar to heart cells. These substances are proteins, and 75 percent

THE CELL IN DEVELOPMENT

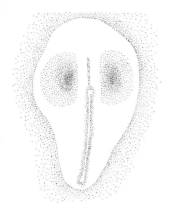

Figure 9.12 **The heart-forming areas of the chick embryo, from which cells can migrate into the site where actual formation takes place. The cells move first toward the tail and then through the region of the primitive streak (stippled lines) into the mesoderm (middle layer of cells), after which they assemble in the head area on either side of the primitive streak. The intensity of the stippling is a measure of the number of cells coming from a given area.** (Courtesy of Dr. Mary E. Rawles.)

of heart protein consists of *actin, myosin,* and *tropomyosin,* the proteins that are necessary for muscle contraction. Heart myosin, however, is different from leg-muscle myosin and can be distinguished from it by chemical tests. Tests for actin and myosin reveal that these proteins are more generally formed by cells earlier in an embryo's life than later (Figure 9.13), and that the myosin appears earlier and more widely spread than does actin, although both are localized later in the same region. It is possible that the localization occurs through the migration of cells, but it is more likely that some cells lose their ability to synthesize myosin, whereas others, in the heart-forming region, have this ability accentuated.

The interesting point is that we can chemically recognize heart cells long before they have either reached the heart region or assumed a shape characteristic of heart-muscle cells. Also, if we interfere with the early synthesis of myosin by giving the cells an inhibitor, antimycin A, formation of the heart can be prevented.

These heart studies emphasize that the differentiation of cells is essentially a change in cellular proteins. Most, if not all, of the morphological features of the cell have proteins in their chemical makeup; a change in these features is undoubtedly a reflection of a previous alteration in the proteins. A change in metabolic function must also involve a protein change, because all reactions in the cell are gov-

Figure 9.13 **The distribution of heart-muscle proteins in three stages of the chick embryo. Heart actin (lightly dotted area) is formed later than heart myosin (heavily dotted area) and in a much more restricted area of the embryo, but myosin formation itself becomes more restricted as the embryo develops. The density of dotting indicates the intensity of myosin formation. The position of the primitive streak is indicated in the center of the embryo.** (Courtesy of Dr. J. Ebert.)

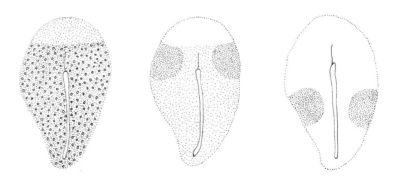

erned by enzymes which themselves are proteins. Since the nuclei of most somatic cells appear to be similar, we must therefore assume that most cellular alterations occurring during differentiation take place in the cytoplasm. The nucleus, however, is the control center of the cell, and we must also assume that cytoplasmic change is preceded by earlier nuclear change of some sort.

The uniqueness of an individual resides in the sequence of nucleotides in its DNA (Chapter 4), which in turn determines the proteins peculiar to that individual. All of the cells of an individual possess a full set of genes; each is, therefore, potentially capable of manufacturing all of the proteins of the body. The obvious fact, however, is that *they do not do so.* Red blood cells produce hemoglobin but nerve cells do not; certain cells of the pancreas produce insulin, others do not; the palisade cells of a plant leaf contain chlorophyll, root cells do not. The protein formation of a cell, by way of transcription and translation, is therefore a function not only of the particular genes it possesses, but also of the environment in which these genes exist. Several examples will make this clear.

1   A plant grown from seed in the dark will not produce chlorophyll. The plastids are present in an undeveloped form, but are nonfunctioning in terms of photosynthesis. The addition of light to the environment will cause the plastids to develop and eventually to begin photosynthesizing. The genes responsible for the formation of chlorophyll are active only in the presence of light. Light, therefore, as a controlling factor, governs the differentiation of the plastids and leads to the selective and coordinated synthesis of a distinctive set of proteins.

2   The bacterium *E. coli* can be grown in a culture medium with the sugar glucose as the only source of carbon and energy; *E. coli* can also be grown in the presence of the sugar galactose. In the presence of glucose, however, the enzyme system required for the utilization of galactose is absent; it appears only when galactose is provided. Galactose is therefore a specific inducer, and is required in order to activate the genes responsible for the enzyme system capable of utilizing galactose.

3   *E. coli,* by a series of chemical reactions, each of which is mediated by a specific enzyme, manufactures its own histidine (an essential amino acid) at a rate appropriate to the needs of the cell. If histidine is added to the culture medium, the genes responsible for the formation of these enzymes are turned off, and remain inactive until the external supply of histidine is exhausted and the necessary enzymes are formed once again. Externally supplied histidine is therefore a repressor, and the cell by means of a feedback repression controls the activity of some of its constituent genes.

The examples just cited are relatively simple demonstrations of a reversible differentiation; the genes are active or inactive according to the environmental circumstances of a nongenetic nature. A more prolonged, but still reversible, differentiation is seen in the sporulation of bacteria. When an adverse environment is met—dehydration or insufficient nutrients—the bacterial cell rounds up, forms a heavy outer wall, and goes into a resting state; no cell division occurs and the rate of metabolism is depressed. The return of favorable conditions reverses the situation.

A more complex situation is encountered in multicellular organisms. The environment for any particular cell also includes the influence of neighboring cells. For example, in the development of the vertebrate eye, the optic cup develops as an outgrowth from the ventrolateral portion of the brain. When this outgrowth reaches the external epidermis, the epidermal cells differentiate into a lens. If the optic cup is removed, the lens fails to form. The presence of the optic cup is therefore necessary as an inducer of differentiation, a point made even clearer by the fact that an optic cup of the right stage transplanted to the back of an animal (a frog) will induce the back epidermis to form a lens, something it would not normally do.

It is also evident that differentiation in multicellular organisms is a progressive affair requiring several generations of cells for complete transformation of an irreversible sort. The discussions related to Figures 9.9, 9.10, and 9.13 indicate this. The inducing and/or repressing influences, if comparable to the *E. coli* situation previously discussed, are not as easily identifiable, yet we assume that similar actions are operative to govern gene activity. We can then ask the following questions: (1) are the selective responses of cells that lead to differentiation initiated in the cytoplasm, and do they in turn have an effect on the genes in the chromosome; or (2) is the chromosome itself modified initially? Several studies suggest that differentiation begins at the chromosomal level.

Man possesses an XX-XY sex-determining chromosomal mechanism, with the female being XX and the male XY. At about the fourteenth day of prenatal life of females, when the embryonic mass consists of several thousands of cells, one of the X chromosomes in all cells except those destined for the germ line becomes transformed into a

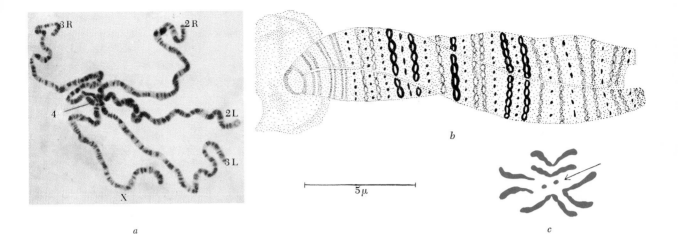

a

b

5μ

c

Figure 9.14   Salivary-gland chromo-
somes of Drosphilia melanogaster. *(a)*
A smear preparation from the salivary
gland of a female, showing the X
chromosome, the arms of the two
autosomes (2L, 2R, 3L, and 3R), and
the small chromosome 4. The diploid
number of chromosomes is present,
but the homologues are in intimate
synapsis and are united by their cen-
tric heterochromatin into a chromo-
center. (b) Enlarged drawing of chro-
mosome 4, showing the banded
structure; the diffuse chromocenter is
at the left, and the two homologues
are intimately paired. (c) Metaphase
chromosomes from a ganglion cell,
with an arrow pointing to the chro-
mosomes 4 and with a scale to indi-
cate differences in size between the
two types of chromosomes. [(a) Cour-
tesy of Dr. B. P. Kaufmann; (b, c)
courtesy of C. B. Bridges, Journal of
Heredity, *26 (1935), 60–64.]*

heterochromatinized, inactive body. This chromosome continues to replicate at each cell division, but it is permanently differentiated into a functionally inactive state. The somatic cells of the male and female are comparable therefore in possessing only one genetically active X chromosome.

On a more selective basis, the same phenomenon can be demonstrated at the genic level through nuclear transplantation. The nucleus of a somatic cell can be sucked out with a micropipette and injected into an egg which has had its own nucleus removed. If done with nuclei from the late blastula or gastrula of a frog, the results are variable: some nuclei are capable of forming complete and well-developed embryos, others show, to varying degrees, that they have lost this capacity (Figure 9.8). The interesting thing is that the capacity for promoting a particular degree of embryonic development is retained through several transplant generations. The lost capacity for full development is not completely irreversible, but irreversibility becomes more pronounced as the cell becomes older and the more differentiated.

We also know that chromosomes other than mammalian X chromosomes often exhibit morphological differentiation. In the larvae of dipteran (single-winged) flies such as *Drosophila melanogaster,* the cells of the salivary glands, in particular, contain very large, banded chromosomes, as shown in Figure 9.14. They arise by the growth and elongation of ordinary chromosomes, and their appearance in tissues other than that of the salivary gland also changes during development (Figure 9.15). These changes, which affect the appear-

THE CELL IN DEVELOPMENT

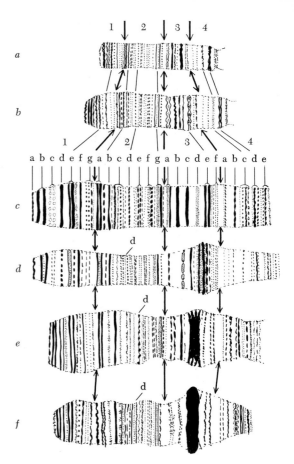

Figure 9.15   The distal end of a salivary-gland chromosome of the dipteran midge, Rhyncosciara angelae, *showing the varied appearance at different stages of development: (a) young larva; (b) 8 days older than (a); (c) full-grown larva; (d–f) successive stages of development between (b) and (c), during which time a number of the bands alter their appearance. An increased dimension of a band indicates that RNA is being formed.*

ance of particular bands in the chromosomes, indicate that the genes in these areas are particularly active at one stage but not at another, and the puffing is a morphological indication of active RNA production. The precise patterns of genic induction and repression remain to be discovered.

INTEGRATION

We can now appreciate that the fertilized egg is a rather remarkable cell. It is primarily an organism in its simplest, or undeveloped, state; it is a cell only secondarily, and differs from other cells in its potentiality for total development. Growth and differentiation, as we have seen, are two of the processes by which development is achieved.

THE CELL

When we consider development in terms of cells, therefore, we find a progression from initially simple and uniform cell types to complex and diverse ones; the plastic, versatile cell becomes stabilized with an unalterable structure; cells with general functions develop highly specialized functions.

But growth, relative growth rates, and differentiation are not enough to account for all of development. The whole course of development of an organism, from the moment of fertilization to death, is characterized by a *unity* and a *harmony* of structure and behavior that cannot be explained only by these processes. The egg, the seed, the embryo, and the larva are all organisms as complete as is the mature individual that arises from them. Although their development may not be fully realized, at all times they behave like, and indeed are, fully functional living entities, and they develop as a *whole*, not simply as a collection or group of cells.

This phenomenon of unity is difficult to define and even more difficult to comprehend. It involves regional controls, the interaction of one kind of cell with another, and the development of form. To understand integration of this sort requires an intimate knowledge of organization and interaction at the molecular, cellular, tissue, and organ levels, and this we do not possess except in a fragmentary way. We do know that integration depends on a number of factors: chemical stimuli such as hormones; cell movements such as those involved in the formation of the heart and gonads and in the infolding that produces the gastrula; cell interactions such as those responsible for the formation of a compound structure such as the eye (Figure 9.5); and the processes of differentiation that, for example, impress upon an undeveloped limb bud the ability to form a complete leg even after the bud is removed from its normal position of development (Figure 9.10).

We can highlight the significance of such integration by posing several questions that we cannot yet answer. Why do organs or organisms reach a mature size and stop growing? What determines the life span of organisms? What determines the size relationship between one part of the body and another? What is it that determines *morphogenesis*, the origin and realization of form? How does one phase of development affect succeeding phases? We must answer these and many other questions before we begin to understand development as an overall process of life. Some of the aspects of development can possibly be approached and explained at the cellular level, but others, such as morphogenesis, seem to involve a higher order of organization in which the individual cell plays a subordinate role. And at a vital, practical level, our ability to control cancer, which results from the growth and escape of cells from regulative control, rests on our comprehending the functioning cell and the processes of development as rigidly governed systems of chemical checks and balances.

ANY INDIVIDUAL HAS A LIFE SPAN THAT IS CHARACTERISTIC OF THE species to which it belongs. In some instances, the life span is rather exactly delimited: 17 years for certain periodic locusts. Usually, however, we think of the life span as an average figure: a few days for certain insects, a few months for annual plants, three score and ten years for man, 250 to 300 years for an oak tree. The redwoods of our West Coast and the bristlecone pine of California's White Mountains are probably the longest lived organisms; some of the trees reach an age of several thousand years.

Cells, too, have a life span that they complete, and after which they die. And, like organisms, particular kinds of cells, even in the same organism, have characteristic long or short life spans. Yet it is entirely reasonable to consider some cells to be immortal. When a unicellular organism divides, the life of the single cell becomes part of the life of two new cells, and as long as the species live so, in a sense, does the original cell; the life of a number of such a species, then, stretches in an unbroken chain back to some original cell in the past. Among sexually reproducing organisms, only the cells of the germ line can lay claim to immortality, for they are the only cells that connect succeeding generations, contribute through division, growth, and differentiation to descendant organisms, and thus keep the species alive. We are all familiar with death as it involves whole organisms, but among the cells of the body, death is a normal and necessary process, because if its role is altered the functioning of the organism will be affected drastically. As a biological problem, and apart from the death of an organism, there are two broad categories of cellular death: (1) that resulting from the wear and tear of existence, which, during much of the life span of an organism, is generally counterbalanced by an equivalent amount of cell replacement; and (2) that resulting from the normal processes of development.

CELL REPLACEMENT

It has been estimated that a human being has a new body every seven years, the time it takes for the old cells of the body to be replaced by new ones. Even if accurate, this figure is very misleading, for some parts of the body require a constant replacement of cells while other parts either require none or are incapable of replacement. By the time of birth, most of the nerve and muscle cells of the body have been formed, and they will continue to function as long as the individual lives (barring injury). If a nerve cell is destroyed, it is not replaced by another, and the nerve cell cannot divide once it is fully differentiated. No general nerve-cell replacement center exists, as is true for blood cells, but there is now some evidence that some small, interconnecting neurons in the human brain are induced by exposure to learning situations at an early age. Recent studies also suggest that the muscles are capable of limited replacement, although the replacement is probably by cells other than those that are fully differentiated. That an organ remains constant in size is not, therefore, indicative of its rate of replacement; unless the cellular conditions are known, the constancy of size merely indicates that there is no net gain or less of cells. The death of the old cells can be equalled by the production of new cells.

Some biologists estimate that each day the human body loses 1 to 2 percent of its cells through death. Body weight, therefore, would double every 50 to 100 days if no cells died, and if cell division proceeded normally. If the body weight remains constant, therefore, dead cells must be replaced by new ones, by billions of cells every day. Since virtually none of these is produced in the muscles or nervous tissue, there must be active centers of death and replacement elsewhere. These would include the protective layers, and regions of the blood-forming, digestive, and reproductive systems. The other organs of the body have much slower replacement rates; a liver cell, for example, has been estimated to have an average life span of about 18 months. Consequently, if we were to look at a slice of liver under a microscope, we would expect to find very few cells in division. On the other hand, if a portion of the human liver is lost through surgery, the rate of cell replacement is stepped up until the original size is once more approximated. The loss of cells in the liver apparently induces the remaining cells to undergo active division.

The outer surface of the human body is covered with a protective layer, which is mostly skin, but which also includes the lining of all openings, the cornea of the eye, and such modified skin derivatives as nails and hair. The cells of these structures are constantly being lost through death: the skin sloughs off, and the growing nails and hair are composed of dead cells. The process of replacement, then, must be a relatively rapid one. The underlying cells are constantly dividing, and are pushed outward toward the skin surface, while the outermost cells become cornified (hardened) as they die (Figure 10.1). It takes approximately 12 to 14 days for a cell in the skin of the forearm to move from the dividing to the outermost layer of the skin. Calluses on the hands are thickened areas of dead cells, and a needle can be pushed through these areas without causing pain or drawing blood.

The cornea of the eye is a special type of skin in which the rate of cell death and replacement is high. The cornea, in fact, is an excellent type of tissue to examine for active cell division. Since it is only a few cell layers thick, it can be stripped off (the salamander and rat are good animals to use for this purpose), fixed, stained, and mounted intact on a microscope slide. The dying cells can be seen at the outer surface, while the cells underneath are in active division.

The cells of the blood are not formed in the blood. The red blood cells are derived from the bone marrow, and the white cells (leucocytes), of which there are several kinds, from the lymph nodes, spleen, and thymus gland as well as from bone marrow. Together, these cells and the *plasma* constitute the blood, which has an average ratio of one white cell to 400 to 500 red cells. When the leucocyte-forming areas overproduce, a form of blood cancer, or leukemia, results: myeloid leukemia if the bone marrow is overactive; lymphatic leukemia when other areas such as the thymus gland are overactive. The blood-forming areas, however, usually manage to maintain the cell ratios, but obviously there must be a high loss of cells to offset the new ones formed or the blood system would clog up. Since each

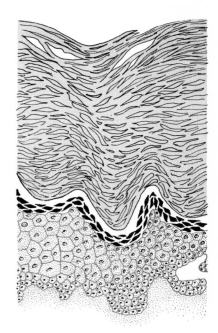

Figure 10.1 **Section through human skin,** *showing the progression of cells from the region of division at the base up to the horny layer of dead cells at the surface. The surface layer is continuously eroded away but is being as continuously replaced by cells moving into place from below.*

THE CELL IN DEATH

type of cell dies off at a relatively constant rate, we need consider here only one, the red blood cells. Their life span is about 120 days. They lack a nucleus and virtually all other cytoplasmic organelles, having lost these as they passed into the blood stream. The wear and tear on the plasma membrane as a result of passage through the vessels cannot be repaired, and the cells grow fragile and finally burst. Certain types of illness may shorten their life span. In a patient with pernicious anemia, the life span is reduced to about 85 days; with sickle-cell anemia, to 42 days. The rate of replacement cannot keep up with the loss of cells, and the red-cell count falls below normal and results in an anemic state. The cause of the shortened life span of these cells is not known, but it is clearly related to the fragility of the cells, and hence to the character of the plasma membrane.

The digestive system is another organ with a cell death rate that is very high. It has been estimated that the cells lining the intestine of the rat are replaced every 38 hr, while the surface epithelium of the stomach is replaced every 3 days. The surface cells are constantly being sloughed off, and are replaced by the division of cells which contain a good deal of mucus. In the human, the epithelial cells of the duodenum are replaced every 1.57 days, and those of the ileum every 1.35 days.

In the plant kingdom, we find that the lower plants—the algae and fungi, in particular—have a rather low loss of cells through death. In the higher plants, however, the rate is enormous. In herbaceous plants, all the cells above ground are lost every season. But consider a large tree. The annual loss of cells in the leaves, flowers, and fruits (only the seeds remain alive) is high enough, but when you add all the cells going to form dead wood and bark, it is readily apparent that the loss of cells through death in animals is small by comparison. Yet a high rate of cell death is as much a normal pattern of existence as the continuation of living cells.

The difference between higher plants and animals, as regards longevity and cell replacement, is even more profound than it appears at first glance. A plant such as a tree preserves its "youthfulness" and vigor in two ways: (1) by continuing the processes of cell division and differentiation indefinitely; (2) by paralleling these processes with the equally continuous death of differentiated cells, either by discarding the cells at the end of the growing season, as in the case of leaves, or by constantly converting them into dead supportive tissue, as in the case of wood. A mammal, on the other hand, achieves its longevity by an exactly opposite process, that is, by preserving the majority of its differentiated cells in a living, functioning state. As was pointed out earlier, however, a cell that does not divide is destined to die; the result is that the life span of a mammal, or indeed any vertebrate, is short compared to that of the longest lived trees.

When we think of normal development, we naturally think of an increase in the number of cells, their subsequent differentiation into specialized cells, and the grouping of these cells into organs and organ systems. This process is dynamic and creative, so it may seem incongruous to characterize cell death as a vital and necessary aspect of development. Cell death, however, plays two very significant roles in development. The first of these, *metamorphosis,* has long been known; the second, the role of cell death in the shaping of organs and body contours, is only beginning to be appreciated and investigated as a phase of development.

Metamorphosis involves a change in shape (the transformation of a larval form of an organism into an adult) and a change in organs when one mode of life is exchanged for another. Two well-known examples are the metamorphosis of a tadpole into a frog, and that of a caterpillar into a pupa and then into a butterfly or a moth.

A tadpole, at the time of metamorphosis, is transformed into a frog without any great change in size, and in the common American leopard frog the process takes about a year. The tadpole that emerges from the egg, and the large tadpole about to metamorphose have the same general shape; in its conversion into a frog, it grows legs and loses its tail, which is devoured by wandering cells, or *phagocytes,* that are carried by the blood stream to the tail region where they gradually consume the muscles, nerves, skin, and other tissues. The skin shrinks and eventually the tail is reduced to a mere stump. In addition, the tissues in the digestive and excretory systems are extensively reorganized. The process can be speeded up or slowed down experimentally, for in the frog, metamorphosis is, at least in part, under the control of an iodine-containing hormone, thyroxine, from the *thyroid gland.* More thyroid hormone accelerates the process, less reduces the rate of change and may even prolong larval life and shape indefinitely. An interesting aspect of such cellular changes in the frog is that the same thyroxine that governs cell death in some tissues is responsible for the growth and differentiation of other tissues contributing to adult form and function. The capacity of a cell to degenerate or to grow actively in response to the hormone must, therefore, be a function of the state of the cell.

The character of metamorphosis in insects varies quite widely, and

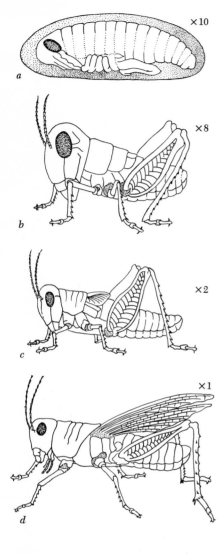

×10

a

×8

b

×2

c

×1

d

Figure 10.2 **Incomplete metamor-phosis as it occurs in a grasshopper. A gradual transformation with an in-crease in size takes place rather than a complete change of form as in the larva-pupa-adult transformation char-acteristic of flies, moths, and butter-flies.**

THE CELL

cell death is not always a major aspect of change. In the simplest type of metamorphosis, the cells of a particular larval tissue are retained and enlarged to form the corresponding tissue in the adult, and only minor differences in growth and differentiation are needed to bring this about. In these cases of *incomplete metamorphosis,* the form of the insect is only slightly altered—for example, the addition of fully developed and functional wings—as the juvenile form matures to adult proportions. Good examples of this type of metamorphosis are found in locusts, grasshoppers, and cockroaches (Figure 10.2).

In *complete metamorphosis,* the larval and adult forms are totally different from each other. The larva, or caterpillar, is converted into a pupa; the larval skin hardens and shrinks into the outer skin, or *puparium,* of the pupa, and the larval tissues are almost completely destroyed. The adult develops during pupation, and adult tissues arise from *imaginal buds* that form in the larva and that escape cell death. These buds can be regarded as zones of persistent embryonic tissue in which the potentiality for growth and differentiation is suppressed during larval life, and is realized only when the *juvenile hormone* (which controls larval growth) lessens its normal activity and *ecdy-sone* (the hormone concerned with molting and metamorphosis) be-comes fully effective.

Figure 10.3 shows the location of certain imaginal buds of the larva of the fruit fly, *Drosophila.* Much of the brain and nervous sys-tem will survive cell destruction, but the intestine, blood system, muscles, and skin will be totally destroyed and then replaced.

The last type of cell death we shall consider is that involved in the shaping of organs. Form can be achieved by relative rates of cell death as well as by relative rates of cell growth. As organs de-velop during morphogenesis, for instance, excess cells are often a hindrance, and these transient cells that are of use to the embryo or larva but not to the adult must be removed (the tail of the tad-pole is a case in point, as is the embryonic tail of humans that is formed, but resorbed, before birth). Or when an organism forms secretory ducts, cells often die instead of pulling apart to provide for the central hole, or lumen. Many organs form by the infolding of tis-sues that then fuse along their edges—for example, the eye and part of the nervous system—and the seams where fusion takes place are removed by cell death. Fingers and toes are separated from one an-other in the same way; if cell death and cell resorption fail to take place, and the separation is incomplete, a webbed condition, or *syn-dactyly,* results.

One of the most arresting instances of cell death as a morpho-genetic phemomenon is the one that frees the elbow of the wing of the chick from the body wall and gives the wing its characteristic shape. Figure 10.4 illustrates the region in question. These cells die as a line of cellular destruction moves from the body area along the front and back of the wing toward its tip, thus separating the elbow region from the body wall, and sculpturing the form of the wing. What is most fascinating of all in this process, however, is that when these cells are removed and transplanted to another part of the embryo, the cells *die on schedule* (at approximately 4 days of age) as if they were still in their original site. Their time of death had already been determined by some unknown change that had taken place within them, and once embarked on their course of destruction, they could not escape. Such programmed cell death is, of course, a specific instance of differentiation, which includes elimination.

We may well ask why cells die, since this seems an extravagant waste of materials that took substance and energy to form. Dying cells are not hard to recognize. The nuclei become compact and dense, the lysosomes and the destructive enzymes in them increase, and macrophages (scavenger cells) move in to dispose of the dying cells. In developing systems, where cell death contributes to the emergence of form, the causes are many. The death of cells in the wing bud is dependent, at least in part, on underlying tissues because if removed, by transplantation, to other regions of the body, they become an integral part of the host site. If placed alone in tissue culture, they grow indefinitely, but if the underlying tissues are placed adjacent to them, they soon die. Some diffusible substance would appear to be the causative agent of death. In the instances of metamorphosis we have mentioned, hormones are involved: a tadpole remains a tadpole if thyroxine is decreased, and a caterpillar remains a caterpillar if juvenile hormone is plentiful and ecdysone is absent. In other instances a genetic influence on cell death is evident. In *Drosophila* some females are sterile because certain genes cause an early death of the nurse cells that normally surround the developing egg; in the mouse, a variety of genes cause selective destruction of parts of the tail; in humans, the absence of cell death, resulting in syndactyly, is an inherited abnormality. In the absence of positive evidence, we can only assume that the nucleus controls cell death as well as the normal living activities of the cell. As pointed out in Chapter 9, development is an extraordinarily precise phenomenon, consisting of many coordinated events and processes. Even though our knowledge of the causes of cell death is meager, we must view such death as part of the normal processes and as important as growth and differentiation.

We know even less of cell death in organisms that die naturally of old age. Old cells look different morphologically from young cells.

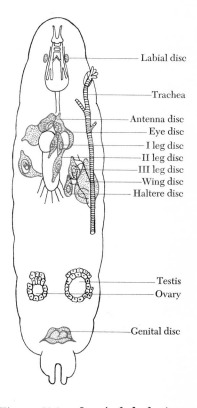

Figure 10.3 *Imaginal buds in a mature larva of* Drosophila. *During metamorphosis most of the larval structures except the nervous system will undergo destruction, while the adult tissues will arise from the imaginal buds, some of which are indicated. The buds form during larval life but do not undergo differentiation until the influence of the larval hormones wanes.* [Reprinted with permission by D. Bodenstein from M. Demerec, *Biology of Drosophila* (New York: John Wiley & Sons, Inc., 1950).]

THE CELL IN DEATH

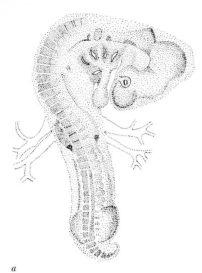

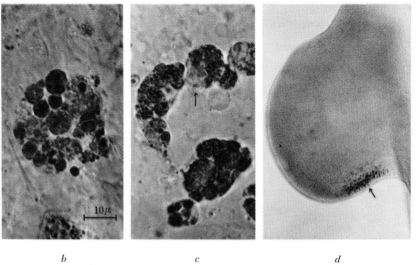

a

b

c

d

Figure 10.4 **The wing bud of a developing chick (a), and the cells in the axial region that die: (b, c) dying cells, which stain heavily; (d) the wing bud, with dying cells stained.** (Courtesy of Dr. John Saunders.)

In humans, aging cells accumulate a pigment that contains proteins and enzymes, but the relation of the pigment to aging is not known. About the only thing we can say at the present time is that to stay young a cell must divide. If it differentiates, it writes its own death sentence.

BLOOM, W., AND D. W. FAWCETT   *A Textbook of Histology*. Philadelphia: W. B. Saunders Co., 1962. Widely used by medical students, this is one of the best American texts available in its field; beautifully illustrated in color and in black and white, and with many electron micrographs.

BUTLER, J. A. V.   *Inside the Living Cell*. New York: Basic Books, Inc., 1959. A nontechnical account of cell structure and function and of the reactions of cells to radiations, chemicals, cancer, and aging.

FAWCETT, D. W.   *The Cell: Its Organelles and Inclusions*. Philadelphia: W. B. Saunders Co., 1966. A superb collection of electron micrographs of vertebrate cells.

GERARD, R. W.   *Unresting Cells*. New York: Harper & Row, Publishers, 1940. One of the finest books available on cells; should be read by every student of biology.

HOFFMAN, J. G.   *The Life and Death of Cells*. New York: Doubleday & Company, Inc., 1957. A semipopular account of the microscopic realm of cells, as viewed through modern theories of matter and energy.

HUGHES, A.   *A History of Cytology*. New York: Abelard-Schuman Limited, 1959. An excellent historical account of microscopical observations, the cell theory, cell division, theories of inheritance, studies of the cytoplasm, and the place of cellular theory in general biology.

JENSEN, W. A.   *The Plant Cell.* Belmont, California: Wadsworth Publishing Co., Inc., 1964. A comprehensive treatment of the cells of higher plants.

MCLEISH, J., AND B. SNOAD   *Looking at Chromosomes.* New York: St. Martin's Press, Inc., 1958. A small book that describes mitotic and meiotic divisions in the lily; illustrated with a superb collection of photographs.

MERCER, E. H.   *Cells: Their Structure and Function.* New York: Doubleday and Co., Inc., 1962. Similar to this volume, but with a more biochemical emphasis.

STERN, H., AND D. L. NANNEY   *The Biology of Cells.* New York: John Wiley & Sons, Inc., 1965. A comprehensive treatment with a strong biochemical emphasis.

STREHLER, B. L.   *Time, Cells, and Aging.* New York: Academic Press, Inc., 1962. An excellent small volume that deals with aging as a cellular problem.

SWANSON, C. P.   *Cytology and Cytogenetics.* Englewood Cliffs, N. J.: Prentice-Hall, Inc., 1957. A detailed account of chromosomal structure and behavior, particularly as these relate to genetics and evolution.

WILSON, E. B.   *The Cell in Development and Heredity.* New York: The Macmillan Company, 1925. A classic of biological literature, with discussions oriented toward embryology; worthwhile for both beginner and expert.